Lockheed PV-1 Ventura in action

by Charles L. Scrivner &
Capt. W.E. Scarborough, USN (Ret.)

illustrated by Don Greer

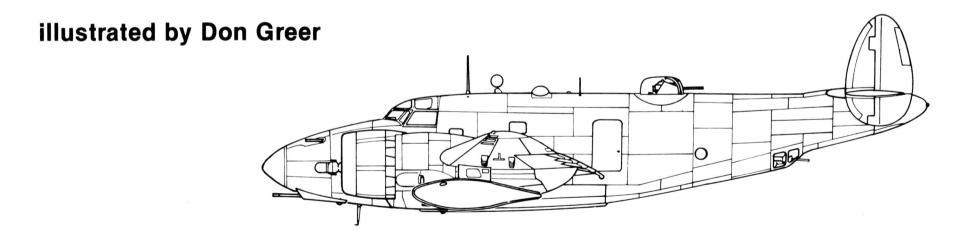

 squadron/signal publications

Cover

VB-136's LT F.R. Littleton shoots down a Tojo in flames over Paramushire, 17 September 1944.

Naval Aviator

Combat Air Crewman

COPYRIGHT © 1981 SQUADRON/SIGNAL PUBLICATIONS, INC.
1115 CROWLEY DRIVE, CARROLLTON, TEXAS 75011-5010
All rights reserved. No part of this publication may be reproduced, stored in a retrival system or transmitted in any form by any means electrical, mechanical or otherwise, without written permission of the publisher.

ISBN 0-89747-118-0

If you have any photographs of the aircraft, armor, soldiers or ships of any nation, particularly wartime snapshots, why not share them with us and help make Squadron/Signal's books all the more interesting and complete in the future. Any photograph sent to us will be copied and the original returned. The donor will be fully credited for any photos used. Please send them to: Squadron/Signal Publications, Inc., 1115 Crowley Dr., Carrollton, TX 75011-5010.

To a truly great airplane — the PV, to the pilots and aircrewmen who flew them — especially those who did not return and to our wives — Florence and Sarah — for their patience in this endeavor.

In the author's opinion, the best in-flight picture of a PV-1 Ventura extant. The PV's graceful lines, batwing silhouette and Lockheed-family tail are evident. The plane is searching for U-boats in the Atlantic with ASD-1 radar scanning. It's 1944 and the PV-1 is in Measure II Anti-Submarine Camouflage — for use only in areas where enemy air opposition was not anticipated. (Lockheed)

First flight of the prototype Ventura I (RAF serial AE658) on 31 July 1941, also in Sand and Spinach camouflage. The Boulton-Paul turret has been installed, but no guns. Note the bubbles on canopy. These were eliminated on US production Venturas. (Lockheed)

Introduction

Although their contribution to the Allied victory in World War II has been grossly neglected by aero-historians, a review of the record reveals that the Lockheed PV-1 Ventura was a dependable, fast, hard-hitting attacker who's abilities exceeded its 'P-for-Patrol' designation. From early 1943 until war's end in 1945, Venturas ranged the vast Pacific - from the Aleutians to Australia and westward to the home islands of the Japanese Empire itself. Hostility toward the Ventura, born of inadequate training during the initial introduction of this demanding high-performance airplane, quickly gave way to full and unqualified acceptance by flight crews and operational commanders. The Ventura's reputation as an airplane that could be depended upon to bring her crews back home was quickly established and continued to grow.

The Ventura was emphatically not, as has been implied in some sources, a warmed-over, enlarged Hudson. The British Purchasing Commission, searching for a replacement for their aging Coastal Command Hudsons, requested Lockheed to undertake the project, basing the new airplane on the commercial Model 18 Lodestar design (as the Hudson had been based upon the smaller Model 14), utilizing the maximum amount of existing tooling. The resemblance between the Hudson and the Ventura was due to their having been designed by the same team.

Specifying the 2000hp P&W GR 2800 Double Wasp engine introduced a design problem, the solution of which caused one of the Ventura's most distinguishing characteristics. The more powerful engines for the new plane required a larger propeller, but the requirement to utilize the existing Model 18 tooling precluded any change in nacelle location. The problem was solved by utilizing a 'paddle-bladed' propeller. The wide blades provided the required additional area with no increase in propeller diameter. These blades, and their proximity to the fuselage, produced a very distinctive sound, most pronounced when a Ventura flew by overhead, and unlike that of other Allied aircraft operational at that time. Also specified by the RAF was a bomb bay capable of holding up to 2500 lbs of payload and eight .303 cal machine guns. The guns were to be distributed to the nose (four), to a Boulton-Paul dorsal turret (two) and two more mounted under the tail as a stinger. While conventional in appearance and construction, the Ventura was unique in being the first US bomber with all-metal control surfaces.

Lockheed received an initial order for 300 of a proposed 675 aircraft buy in May 1940. The first flight of the new plane, designated Ventura I, came on 31 July 1941. Production had been assigned by Lockheed to its subsidiary, Vega, also located at Burbank, California. This accounts for the name assigned to the new aircraft by the British. (Aircraft name policy dictated that names should be alliterative with that of the manufacturer. Ventura -freely translated as Lucky Star - was the choice for the new bomber.) When deliveries began in 1942, a shortage of offensive aircraft in the RAF caused a change in plans and the first Venturas were assigned to Bomber Command's No. 21 Squadron in October 1942, rather than Coastal Command. The first major combat utilization of the Ventura was in an abortive strike against targets in Holland by three squadrons of 2 Group (No. 21 RAF, No. 464 SAAF and No. 487 RNZAF) on 6 December. Of 25 Venturas from 464 and 487 squadrons, 13 failed to return from the daylight low-level attack on the heavily defended industrial plants at Enidhaven.

After 188 Ventura Is had been delivered to the RAF, several changes were made in design and the designation became Ventura II. Those changes included the adoption of the somewhat more powerful '-31' version of the Double Wasp. An additional 200 were ordered under the designation Ventura IIA, which now carried the USAAF designation B-34.

VP-82's PBO-1 Hudson is still in RAF Sand and Spinach camouflage and carries early 1942 US insignia. British radio and armament, including the Boulton-Paul turret were retained. (Bowers)

B-37 Waist Gun Position

The B-37 was the basic B-34 with Wright R-2600 engines. The AAF accepted only eighteen B-37s and made no operational use of them, using them only as trainers and target tugs. A recessed mount for a flexible .30 cal gun on the sides of the fuselage was the most noticeable external difference between B-34 and B-37. (Lockheed)

The RAF quickly became disenchanted with their new bomber. While fast (top speed 314mph), the Ventura never should have been used for low-level attacks on well-defended targets without considerably more experience in the aircraft. The biggest problem, as far as the RAF was concerned, was the fact that the Ventura was a 'hot' aircraft. The Ventura was fast and had high wing loading with the inevitable lack of 'forgiveness' for pilot errors. No more than 300 Ventura IIs were delivered to the RAF, which put them to use as a general reconnaissance aircraft.

This loss of interest by the RAF didn't have a serious effect on Ventura production. Looking for an interim medium bomber to fill an acute shortage of this type in its inventory, the USAAF had requisitioned 208 machines off the Vega production line in 1941. The first was accepted in September, being given the serial number 41-38020. Since they had not been built to US specifications, they did not qualify for a USAAF designation in the 'B-for-bomber' series and were, instead, operated under the Lockheed Model No. 37 and their RAF serial numbers. These Model 37 Venturas subsequently picked up an 'R' prefix, becoming R Model 37s to indicate a restricted/obsolete status. The armament of the B-34 differed from the Ventura II. .50 cal Brownings replaced the two fixed .303s in the forward fuselage. Likewise a Martin turret replaced the Boulton-Paul, again mounting .50s in the place of .303s. The waist-position guns were deleted. These aircraft, and a subsequent batch of 200 built to Army specs and designated B-34 (and later RB-34), served initially as coastal patrol aircraft, then as bomber, gunnery and navigation trainers and finally as target tugs. A later Army order for 550 B-34B navigation trainers, with Wright Cyclone R-2600 engines, was cancelled, then reinstated for the same aircraft as O-56 long range observation planes and then cancelled again. The 18 aircraft on the assembly line were completed and delivered as B-37s. Externally, these were nearly identical to the B-34, with two additional .30 cal machine guns aft, mounted in recessed positions in the fuselage sides.

In July 1942, the US Navy completed an agreement with the USAAF to acquire a quantity of Venturas as a land-based supplement to the combat-vulnerable PBY Catalinas, which had been suffering heavy losses during offensive operations early in the war. The Navy had long recognized a need for high-speed, land-based patrol planes, and had proved the point with the PBO Hudsons being operated over the North Atlantic by VP-82. Such an aircraft could search an area in minimum time, while carrying sufficient offensive armament to assure effective attacks against targets-of-opportunity during anti-submarine patrols. The Ventura, with its speed, defensive firepower and ability to survive in an active combat environment, was also seen as useful for reconnaissance and interdiction strikes against enemy bases and supply convoys.

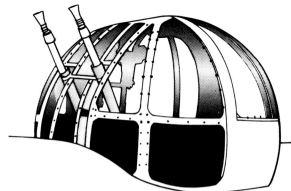

Boulton-Paul Turret

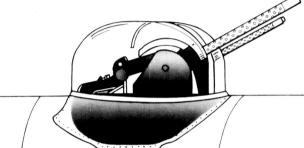

Martin 250CE-13 Turret

PV-3 without Turret

The PV's top turret was undoubtedly one of the best defensive installations on any Allied aircraft of WWII. The Martin 250CE-13 electric turret was a superb gun platform and was equipped with a gyro-computing gunsight. The turret operated from the PV's 24 volt electrical system and was provided with a fire-interrupter to prevent an overzealous gunner from shooting the wings or tail off his own plane during the heat of battle. Ammunition capacity was 400 rounds per gun. Guns were changed manually and fired electrically by MK7 backplate solenoids. Half-inch armor plate extended around the front of the turret from the gunner's ankles to his face. On Tarawa AOM2c Warren B. Herrick of VB-142 is inside the turret. (Warren B. Herrick)

PV-1 Ventura

The Navy Venturas were designated PV-1 ('P' for Patrol, 'V' for Vega). To provide examples of the new plane for training and familiarization while production was getting underway, 27 Ventura IIs were requisitioned from British orders in September 1942, being put into service as PV-3s with British equipment still installed. VP-82, the North Atlantic PBO squadron, was the first Navy Ventura command, accepting a full complement of the PV-3s at NAS Quonset Pt., RI in October 1942.

In late 1942, the Navy assumed all contract administrative responsiblities for production of the PVs at the Lockheed-Vega facilities at Burbank. All Venturas delivered subsequent to this action were assigned US Navy serial numbers and were outfitted with Navy equipment. This included the final 300-plus aircraft from the RAF contract (known as Ventura GR.Mk.V). Some 1600 PV-1s were delivered to the US Navy between December 1942 and the end of production in May 1944.

The PV-1 was essentially similar to the B-34 except for the switch to Navy electronics. The most noticeable change was the replacement of the plexiglass bombadier's nose of the RAF Ventura with a housing of similar (but not identical) shape for the PV-1's ASD-1 radar. This necessitated the deletion of the two nose mounted .30 cal mgs. Not noticeable was the increased fuel tankage which increased the combat radius of the PV-1 to 1660 miles.

The original Lodestar's wing-fuselage geometry and fuselage lines dictated a very shallow bomb bay and restricted the number of bomb stations to six. Payload, weight and balance considerations limited the bomb load to a maximum of 3,000 pounds in the bomb bay. Any mix of bomb type and weight up to that limit could be carried. Typical bomb loading for sea patrol flights consisted of 325 lb or 650 lb depth bombs, for strikes on land bases, 100 lb, 500 lb or 1,000 lb bombs could be carried.

By utilizing an Army-type D-6 shackle, a 2,127 lb Mk 13 aircraft torpedo could be carried in the center of the bomb bay. The PV-1 was also cleared to carry the Tiny Tim aircraft rocket — a 12" diameter weapon grossing some 1,200 pounds. The rocket was carried in the bomb bay on a modified torpedo rack. The Tiny Tim rocket was developed specifically to attack the massive Japanese defensive positions which had resisted, with considerable success, other available attack weapons. Available records, however, do not indicate that PVs made any torpedo or Tiny Tim attacks during World War II.

A 280 gallon auxiliary fuel tank could be carried in the aft half of the bomb bay and another 200 gallon tank in the forward half. PV squadrons in the Aleutians usually carried the aft tank and three 500 lb GP bombs on the forward shackles. On daylight photo-recon missions, both bomb bay tanks were usually carried. This fuel load gave the PVs sufficient fuel for high power/high speed runs over a target with plenty of reserve left for the return flight. The bomb bay tanks were not self-sealing and standard procedure was to empty them first, to reduce the hazard of fire.

The wing pylons, where 165 gallon drop tanks were usually carried, could also carry any type bomb or depth bomb up to 1,000 lb. Thus, the maximum bomb load for a PV on short missions was 5,000 lb. This bomb load, a five man crew and full internal fuel gave the PV-1 a gross take-off weight of 33,200 lb, well above the recommended maximum overload weight of 31,000 lb. Squadrons in combat frequently operated at or above 32,000 lb from short (4,000ft), rough (packed coral) and hot (80° to 90°) runways without difficulty.

The standard Ventura flight crew was five men, two officer pilots and three enlisted aircrewmen, but crew composition could vary with the squadron's mission. Most Pacific squadrons employed an additional crewman, a navigator, most of whom were enlisted men. VB-135, during their second deployment to the Aleutians in 1944, utilized two navigators in each crew, an officer and an enlisted man! Officer navigators were usually pilots, but there were also some non-pilot commissioned navigators in service. The enlisted navigators were qualified aircrewmen (mechanics-AMM and ordnancemen-AOM) who were given special additional training in aerial navigation. The Navy was particularly impressed with the dependability and accuracy of these enlisted navigators. On photo

(Above & Below) VP-82 received the first PV-3s at Quonset Pt. in October 1942, and carrying out operations from Argentia and Quonset Pt. until PV-1s were available in December. VP-93 received PV-3s in November, exchanging them for PV-1s in January 1943. The patch riveted over the turret opening was the most distinguishing external characteristic of the few PV-3s built. (National Archives)

(Below) A prospective Navy crew is introduced to the Ventura on the Vega production line at Burbank. These are early Navy PV-1 production models with the original bomber nose. Each of these early Venturas received a cartoon along with the two color (Blue Gray and Light Gray) color scheme. (Lockheed)

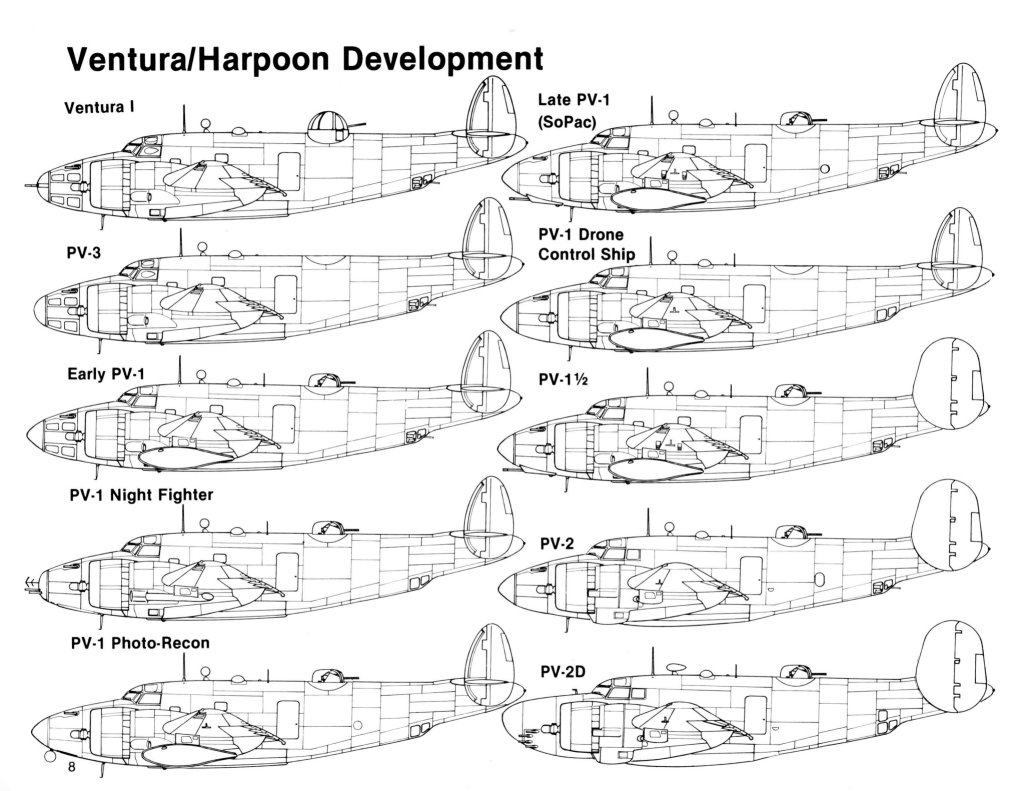

An early Navy production PV-1 at Burbank 1943. Early PV-1s retained the windows on the sides of the nose, but the nose cone was more pointed and was doped over the ADS-1 radar antenna being installed there. The Disney Studios authorized Lockheed to use well-known cartoon characters and many early Venturas, both British and American, were delivered and flew into combat with paintings of Donald Duck and other characters on the side of the rear fuselage. (Lockheed)

Nose Development

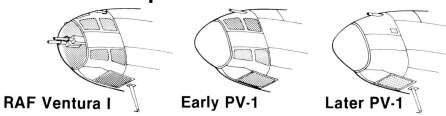

RAF Ventura I   Early PV-1   Later PV-1

recon missions, when high and low oblique and vertical cameras were installed, a Photographer's Mate (PhoM) might be added to the crew. Similarly, an Aerographer might be carried on a weather recon mission. These specialists were not regular flight crew members.

The PV first appeared in the Pacific at opposite poles of that ocean with VB-135's deployment to the Aleutians in April 1943, VB-137's to Wallis Is. in May and VB-140's to Guadalcanal in September. In numerous encounters in the Pacific, Venturas regularly outran enemy fighter planes. Or, if the PV couldn't outrun them, it would shoot them down! This was a unique accomplishment for an airplane classed as a medium bomber, made possible by the fixed guns installed in the nose and the skill of the pilots who flew their PVs like fighters! About a dozen Japanese aircraft fell victim to the Ventura's bow guns in the Pacific. Another dozen were credited to PVs operated by the Marines as night fighters over the Solomons. Other action in the fighter role included long range escort for Army B-24s in the Aleutians and for C-47s on paratroop missions into New Guinea. In a ComAirPac bulletin, No. 21-1943 issued to all commanders, the PV's performance was compared to some enemy fighters:

Zero: The PV-1 can outrun any kind of Zero (Zeke or Hamp), and the floatplane Rufe, at sea level which is where the PVs normally operate. Pilots have had actual experiences in out-running Zekes and Hamps and in one instance, a Hamp was left behind when the encounter was close enough to the enemy home base for the Japanese plane to use full power without danger of running out of fuel.

Tony: The PV has outrun Tonys in combat, but to do so, it needs considerably more than normal rated power.

The twin-row, 18 cylinder P&W R2800s were fabulous examples of reliable, precision machinery. They could absorb an unbelievable amount of battle damage and continue to run. On long over-water flights, in an airplane which would usually float no more than 30 seconds after ditching, the PV crews regularly gambled their lives on the big radials. Squadron combat reports contain numerous examples of the R2800's ability to continue operating after sustaining major battle damage. This quote from a VB-144 report is an example:

> The plane was hit by 20mm explosive shells in the leading edge of the starboard wing, in the fuselage abaft the stinger gun position and in the port engine. The shell which hit the port engine was from the port beam. It exploded on impact with the engine cowling leaving a 10 inch hole. The hit was at the No. 6 cylinder, knocking off the valve cover and housing and causing a bad oil leak.
>
> With the port engine losing oil, no flight instruments, no port rudder cable and bomb bay doors open, the plane flew the 120 miles to Majuro under instrument conditions...maintaining visual contact with plane No. 360. By pumping oil from the auxiliary tank each time the pressure got below 40psi, the pilot was able to use the port engine all the way to Majuro.
>
> This is another example of the stamina of the PV-1 and its ability to take punishment and still bring the crew safely back to base. In particular, it is an additional testimonial to the Pratt & Whitney R2800-31 engine. This is the third instance in this squadron where an R2800 has sustained a direct hit and been seriously damaged by AA fire, but still was able to function and give good performance, thus enabling pilots and crews in all cases to return safely to base.

In this case, had the engine failed, the plane would certainly have been lost since the pilot had no rudder control and would have been unable to compensate for the loss of the engine.

All Venturas were delivered in single-pilot cockpit configuration, the standard British practice in attack bombers. Copilot controls were added by the Navy as soon as they were delivered as a field modification. This is BuNo 48820, the 1170th of the 1600 produced for the Navy. (Lockheed)

Tunnel Gun Position

Interior

Exterior

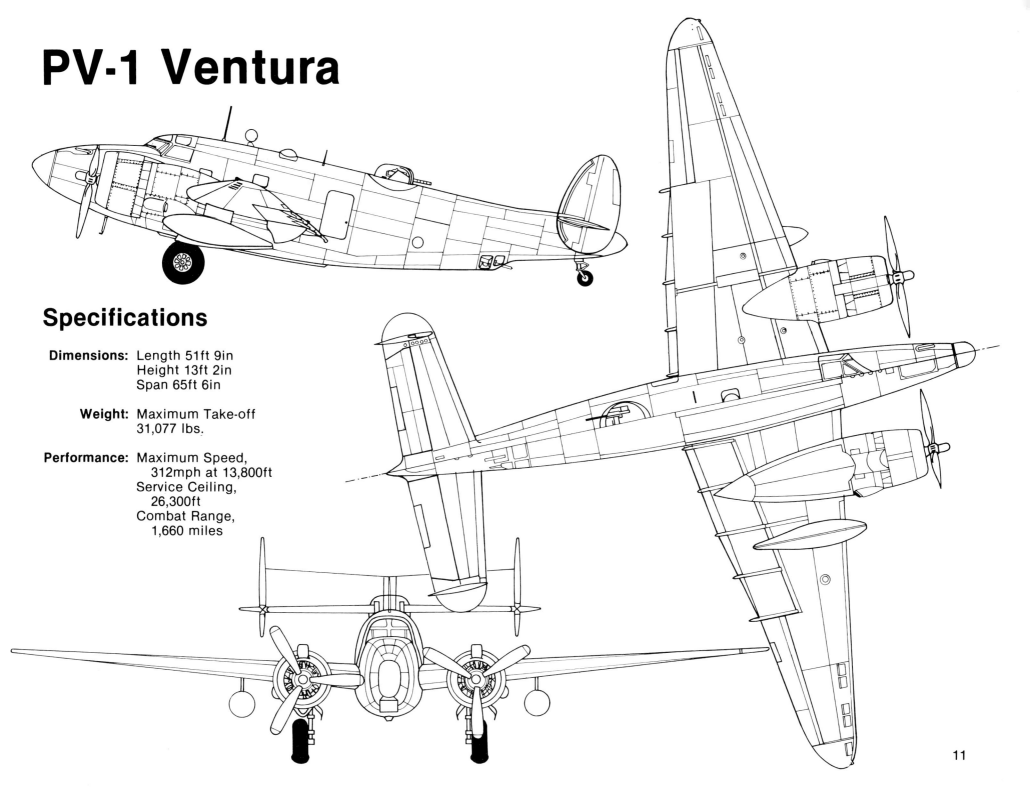

PV-1 Ventura

Specifications

Dimensions: Length 51ft 9in
Height 13ft 2in
Span 65ft 6in

Weight: Maximum Take-off 31,077 lbs.

Performance: Maximum Speed, 312mph at 13,800ft
Service Ceiling, 26,300ft
Combat Range, 1,660 miles

PV-1 on the flight line at NOTS Inyokern, CA, in late 1944 armed with eight 5" HVAR rockets. Note the special camera installation under the port wing just outboard of the rockets. (Authors)

A Whidbey Island based PV-1 drops a Mk 13 torpedo on a practice run. To assure acceptable water-entry angles, aircraft torpedoes were equipped with box-like auxiliary tail fins made from plywood, which broke away on impact. The white plume from the torpedo tail is exhaust, indicating that the motor has already started. (National Archives)

PV-1 Tiny Tim torpedo bay installation, utilizing modified torpedo carrier for a single round. The tail fins protruded through the closed bomb bay doors but had no significant effect on performance of the PV.

The chin gun modification installed under the nose, replacing the flat glass panel, increased a PV's forward firing guns from two to five .50 cals. PVs spat out a lethal dose of fire which exceeded that of some fighters. All five guns were boresighted to converge at 1200 feet and, when carried out by an experienced pilot, a strafing attack could cut a small ship in half. (Lockheed)

(Above Right) A turret gunner (Aviation Ordnanceman) seals the muzzles of his PV's .50 cals with a protective paper cap. (National Archives)

Loading a torpedo aboard a PV. The plywood nose cap improved the torpedo's flight attitude during the free fall and cushioned the water entry as it broke away. (National Archives)

The PV-1 presented an unmistakeable 'bat-wing' silhouette and the tail configuration that identifies it as belonging to the Lockheed family. This early Atlantic-based PV-1 has the unique eight-position national insignias that were a carry-over from pre-war 'Neutrality Patrol' days. (Lockheed)

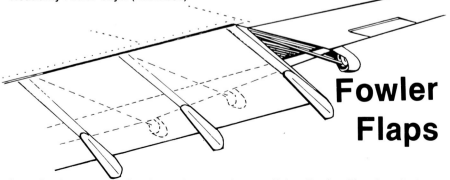

Fowler Flaps

Seen banking away, this later Ventura shows off its Fowler Flap installation. (Lockheed)

The Battle of the Atlantic

The success enjoyed by VP-82 with the PBO Hudson vindicated the Navy's land-based patrol plane concept and supported the action already underway to secure more and better landplanes. To expedite the introduction of the Ventura, 27 Ventura IIs were requisitioned from the RAF production line at Burbank with deliveries to VP-82 being made in October 1942. With their PBO experience, the squadron had no difficulty in transitioning to the Ventura and was soon back at its base, NAS Argentia, providing ASW coverage to North Atlantic convoys. Although they had had some early successes with the PBO, what followed now was a lengthy period of the kind of flying only too familiar to the patrol crews - endless hours of searching for the one whitecap which would suddenly resolve into a periscope plume! The first PV-1 success came on 27 April 1943 when U-174 was sunk by LT(JG) Thomas Kinaszczuk and his crew in the convoy lanes off Newfoundland. In May 1943, the squadron, now designated VB-125, moved to NAS Quonset Pt. for a one month stay, then on to Boca Chica, FL, operating from there and from San Julian until February 1945. The squadron then moved to Natal, Brazil, where it remained under FAW-16 until VE Day.

Meanwhile, the transition to the PVs continued with VP-93 exchanging PBY-5A Catalinas for PV-3s in November 1942 and the British-equipped -3s for 15 new PV-1s in January 1943. In March, VP-93 was redesignated VB-126. As the squadron settled down with the new aircraft, detachments were deployed to Greenland, Argentia and Brunswick, ME, and, later in the year, to MCAS Cherry Pt., NC, and Floyd Bennett Field, NY. A highlight of the period was an attack on a U-boat on 7 August by LT(JG) J.R. Smith and his crew from the Cherry Pt. detachment. The attack was assessed as probable damage to the sub. Later in the year, the squadron was also transferred to FAW-16 and operated from Natal.

As Navy-production Venturas became available during early 1943, new squadrons were commissioned in quick succession. Most of the East Coast squadrons were formed at NAS DeLand, FL, under the operational control of FAW-12. The first of these (the first Ventura squadron to receive the VB designation) was VB-127, commissioned on 1 February 1943. VB-127 completed operational training in May and was deployed to Brazil to join FAW-16. The most significant accomplishment of VB-127's South Atlantic deployment was the confirmed sinking of U-591 off the Brazilian coast on 30 July 1943. In September, VB-127 was transferred to FAW-15 at Port Lyautey, French Morocco. Squadrons based there were responsible for the protection of shipping in the vital convoy lanes leading to the Straits of Gibraltar. A special assignment in October gave VB-127 and its Venturas a chance to prove their offensive capabilities. FAW-15 Catalinas flying patrols through the Canary Islands, over international waters, were being intercepted and fired on by Spanish fighters. Two VB-127 PV-1s were assigned to the patrol on 28 October and, at a point about 7 miles offshore, were jumped by a pair of fighters. Turning into the attack, the PVs opened fire. The startled fighters turned and fled the area, both making forced landings on the nearest beach. Needless to say, FAW-15 patrols were not again harassed by Spanish aircraft. 'Bombing 127' scored again in a highly effective coordinated action on 24 February 1944 when U-761 was sunk off Gibraltar. VB-127 continued ASW operations from Port Lyautey until the end of hostilities.

VB-128 was commissioned at NAS DeLand by FAW-12 on 15 February 1943. As was typical of most VB squadrons forming at this time, few of the pilots had multi-engine experience. Most came from instructor duties or from shipboard OS2U or SOC squadrons. After intensive transition training, these pilots checked-out in the PV. On completion of operational training in mid-May, VB-128 sent a detachment of seven aircraft to Guantanamo Bay, Cuba, to fly convoy escort and ASW patrols as a final shake-down. The remainder of the squadron proceeded to NAS Floyd Bennett Field, and later in the month the detachment from Cuba joined them there. The squadron's first loss occurred on the flight from Boca Chica when one of the Venturas crashed on takeoff following an engine failure. Only the copilot survived that accident. VB-128 operations from Floyd Bennett provided day and night cover to the heavily traveled sea lanes around New York City. No significant contacts were made by the squadron until 7 August 1943 when a submarine was spotted 300 miles off Norfolk. The ready crew, headed by LT(JG) Cross, was ordered to investigate the contact and they took off at 0430. Radar contact was made as they approached the reported location, and Cross turned toward the target for an attack. As the PV emerged from the clouds, it was hit by intense and accurate flak from the sub, both Cross and his copilot were seriously wounded. Despite his injuries, a cockpit full of smoke and an engine out, Cross continued the attack and dropped his depth bombs. The Ventura flew clear of the attack area but was forced to ditch. After a good landing, both pilots and the radioman were able to escape the sinking plane, although the gunner and the plane captain did not. LT Cross died of his wounds in the water, but the other two survivors were picked up by a PBM Dumbo. Cross was posthumously awarded the Navy Cross and his copilot the DFC. On receiving a report that Cross was down at sea, the squadron dispatched a second Ventura, piloted by LT J.M. George. George reached the contact area and reported on station but was not heard from again. An extensive search failed to find any trace of the plane or crew, nor was contact with the U-boat regained.

In August, VB-128 proceeded to Iceland via Quonset Pt., Labrador and Greenland. One plane was forced down enroute, but the crew was uninjured and rejoined the squadron after picking up a replacement PV. The Battle of the Atlantic was now in full swing, and VB-128 joined a British Liberator squadron at Rekjavik. On 3 October, PVs from VB-128

'Sighted Sub Sank Same' Don Mason and his crew with a VB-125 Ventura at Argentia in March of 1943 after the sinking of U-503. Mason had been promoted to LTJG by the time this photo was taken. Left to right, they are: Mason, Copilot M.A. Rigdon, AP1c; Radioman C.D. Mellinger, ARM1c; Plane Captain AAM2c J.J. Nagle and Gunner AMM3c J.A. Holt. (National Archives)

This PV-1, based at NAS Lake City, FL, is coming in for a full-flap landing. The PV's big Fowler Flaps gave the airplane an acceptable landing speed even though it had one of the highest wing loadings (about 65 lbs/sq ft) of any WWII aircraft. Lockheed was firmly committed to the Fowler Flap system and utilized it on most of the company's designs. (National Archives)

Trinidad. The squadron was transferred to FAW-16 in August, moving to Recife, then to Fortaleza. In May 1944, it returned to the States for reorganization at Norfolk and Quonset Pt. then reported to Elizabeth City, where it remained until war's end.

VB-131, following commissioning on 18 March 1943 at DeLand, completed aircrew and weapons training and then deployed to Guantanamo Bay for ASW operations in the Caribbean under the operational control of FAW-11. The Squadron returned to Norfolk in March 1944 for rehabilitation and transfer to the Pacific. 'Bombing 132', also commissioned at DeLand in March 1943, was deployed to French Morocco via NAS Quonset Pt., reporting to FAW-15 at Port Lyautey. It operated there until November 1944, when it returned to Norfolk, then served at Boca Chica until VE Day. VB-132's skipper was LCDR Thomas H. Moorer, who was destined to become Chairman of the Joint Chiefs of Staff in 1970. In 1972, Admiral Moorer became the 'Gray Eagle' - the Navy pilot on active duty with the earliest date of designation as a Naval Aviator. The order of the Gray Eagle is presented "in recogniton of a clear eye, a stout heart, a steady hand, and a daring defiance of gravity and the law of averages".

After commissioning at DeLand on 20 March 1943, VB-133 completed training in July and deployed to San Juan for ASW operations under FAW-11. Within a week of arrival, on 24 July, the squadron had an opportunity to put its training to use. A PAA plane reported a surfaced submarine in the area and the squadron "ready plane", manned by LT R.B. Johnson and his crew, was ordered to investigate. In the excitement of this first combat experience, Johnson forgot to open the bomb bay doors on the first run at the target. Realizing his mistake, he opened the doors and turned quickly for another attack, dropping all six depth bombs on the crash-diving submarine. Post-flight analysis of the crew's reports credited the attack with 'Probable Damage'. On 8 November, the squadron CO, LCDR Murphy, got his chance, but his Ventura was hit by 4-20mm shells from the sub which caused major damage to the port engine but no injuries to the crew. Murphy was able to limp back to base. On 15 November, the squadron was transferrred to Trinidad, sending a 3-plane detachment on to British Guiana. A detachment also operated from Curacao after December. In February, all detachments were recalled to Trinidad and operations continued there until mid-April, when the squadron was ordered to Norfolk for reorganization and transfer to the Pacific.

On 1 April 1943, VB-134 was commissioned at DeLand and began the typical PV squadron organization and training cycle. On receipt of its full complement of Venturas, the squadron moved to MCAS Cherry Point, NC, and operated there under FAW-5 until November. VB-134 joined FAW-16 at Recife and operated there until April 1945.

'Bombing 141' was commissioned at DeLand on 1 June 1943 and joined FAW-11 at Guantanamo Bay on completion of their training. Detachments were sent to San Juan, Trinidad and the Guianas during October and November. In December, the squadron moved to Curacao. The primary mission was ASW, but the squadron also flew patrols designed to intercept Axis blockade runners carrying vital raw materials to Europe. From mid-April to mid-June, crews were rotated to the FAW-5 rocket training course at Boca Chica. In July the squadron was transferred to NAS Beaufort, SC, operating there until VE Day.

VB-143's story differs markedly from that of the other Atlantic squadrons. After commissioning at Boca Chica, the squadron completed training on 16 August. It was ordered to Recife, reporting to FAW-16. On 28 January 1944 it began operations from Ipitanga, near Bahia, Brazil. Missions flown there included coordinated ASW patrols with PBM squadrons VP-204 and VP-211 and ZP-42, a blimp squadron. During this period, U-boat activity in the South Atlantic reached its peak. These squadrons were fully committed, providing day and night convoy escort, barrier patrols and investigating reported sub sightings and RDF contacts. In May 1944, VB-143 moved to Curacao carrying out coordinated operations with VS-37, which was flying the SBD Dauntless. In June, VB-143 was ordered to Boca Chica for rocket and advanced ASW tactics training. On completion, the squadron was split, with six crews proceeding to NAS Chincoteague, VA, for transition training in the PB4Y-1. The remainder continued to operate at Boca Chica as a stand-by ASW unit. As they were relieved by crews from other squadrons, the remaining crews proceeded to Chincoteague. When all crews had completed transition, the squadron, by then redesignated VPB-143, transferred to the Pacific.

piloted by LTs R.D. Bonnell and C.R. Parent attacked and damaged a surfaced U-boat. On the 4th, CDR Westhofen contacted another, which dived before he could make an attack. Westhofen left the scene but returned an hour later, surprising the U-boat on the surface. Despite heavy flak, Westhofen pressed on with his attack and dropped three depth bombs along the length of the hull. The PV crew watched the submarine sink and observed many survivors in the water. The U-boat was later identified as U-336. The squadron moved to San Juan, PR, in December 1943. ASW was still the primary mission, but ideal flying conditions permitted an intensive training program to upgrade pilots and to practice rocket attacks. In June 1944, VB-128 returned to the US, re-equipped with new PV-1s and transferred to the Pacific.

VB-129 was commisssioned at DeLand on 15 February 1943. After completing training, the squadron reported to FAW-16 at Recife, Brazil, in May and operated from Natal, Fortaleza, Recife and Ipitanga. On 11 August, VB-129 participated in a joint action which resulted in the sinking of U-604. In January 1944, the squadron returned to the US, to NAS Quonset Pt. It was moved again to Elizabeth City, NC flying ASW missions from there. A detachment deployed to NAS Brunswick, Maine in January 1945. In March, that detachment returned to Elizabeth City, routine operations continuing until VE Day.

On 1 March 1943, VB-130 joined the other PV squadrons in training at DeLand and Boca Chica. In June, it reported to FAW-11 at San Juan, then, in July, proceeded to Trinidad. On 6 August, in a joint actjon with a PBM and USAAF planes, VB-130 PVs sank U-615 off

'Bombing 145', after commissioning at DeLand on 15 June 1943, completed the training there and at Boca Chica, then proceeded directly to Natal, for duty with FAW-16. The squadron operated continuously from Natal until February 1945, when it moved to San Juan, then to Brunswick, ME, where it remained until the end of hostilities.

The last of the Ventura squadrons commissioned on the East Coast were VB-147 and VB-149, both being activated at NAS Beaufort, SC, VB-147 was commissioned on 15 August 1943 and VB-149 on 1 October 1943. 'Bombing 147' had outfitted and trained at Beaufort, then moved to NAS Floyd Bennett Field in October for operations. After a brief stay at Quonset Pt. in February, the squadron moved to Elizabeth City, NC, in March 1944, then to Trinidad in May and on to San Juan in July. The squadron made its final move to Curacao in December, remaining there until VE Day. VB-149 trained at Beaufort and Boca Chica, then was ordered to MCAS Cherry Point in November 1943. In January 1944, the squadron was moved back to Beaufort and, in July, to the USAAF base at Otis Field, MA. In October, after a brief stay in Quonset Point, VB-149 was ordered to the Pacific.

The 15 Ventura squadrons organized and trained in the Atlantic operated from a variety of bases, ranging from the sophistication of NAS Floyd Bennett Field, in the suburbs of New York City, to nearly completed fields hacked from the jungles of Brazil. The urgency of the situation in the Atlantic, where Nazi U-boat wolfpacks were winning the battle, forced the Navy to expedite in every possible way the training and outfitting of VB squadrons. Shortcuts inevitably led to operational problems - during one month, October 1943, VB-133 experienced 42 complete or partial engine failures in flight! Training, experience and better support solved such problems and, in the end, the Ventura acquitted itself well. The record cannot show the number of convoys saved from attack, but it does clearly show the confirmed U-boat sinkings credited to the Venturas - six between 27 April 1943 and VE Day.

LTJG Thomas Kinaszczuk and his crew from VB-125 with the PV-1 they flew during that attack on U-174. Left to right: Pilot Kinaszczuk, Copilot Lt. R.J. Slagle, Gunner J.A. Holt, Radioman, R.W. Berg. Gunner Holt was flying with Don Mason when the U-503 was sunk. (National Archives)

A mixed bag of PV-1 Venturas on the flight line, Fleet Air Operational Training Squadron VB2-1, at Naval Air Station, Beaufort, SC, May 1944. Shortly after this photo was taken, the squadron accepted delivery of PV-2C HARPOONS as replacement for their Venturas. Note the Martin JM-1 (B-26 with all armament removed, painted "Trainer" yellow, and used as target tugs) in background. (USN via F.A. Henninger)

A typical PV crew represented a potpourri of American origins. Front row left to right: co-author Scrivner (Missouri), Loesner (New Jersey), Breman (Iowa). Back row: Ross (Michigan), Lacey (Kentucky). (Authors)

An Atlantic-based PV-1 carrying the Scheme II ASW paint. Effective October 1944, all Navy aircraft operating in areas where enemy opposition was not anticipated and overcast conditions prevailed were to be painted in Scheme II, which consisted of top surfaces being painted Dark Gull Gray, side surfaces Non-specular White and bottom surfaces Gloss White. (Lockheed)

RCAF, exUSN, PV-1 Ventura. Note RCAF roundel applied over US national insignia on fuselage. This Ventura carries an AI radar antenna alongside its ASD radome.

Central, South and Southwest Pacific

PV squadrons slated for deployment to the Pacific (CenPac, SoPac and SoWesPac) were commissioned and received initial pilot and crew training at West Coast Naval Air Stations Alameda, Moffett, Whidbey and their satellites. After completion of training squadron personnel and their aircraft were loaded aboard aircraft carriers for transportation to Hawaii. After unloading at Pearl Harbor, the planes were flown across Oahu to NAS Kaneohe Bay, headquarters for FAW-2. There a final operational training syllabus, emphasizing ASW, bombing and gunnery, was flown. Usually, as a final 'shakedown' prior to departure for combat areas, three-to-six plane detachments were deployed to Midway and Johnston Islands. VB-137, the first PV-1 squadron in the South Pacific, was deployed to Samoa and the Ellice Islands in May 1943, with detachments being sent to Wallis Is. and Funafuti. The squadron shared patrol and ASW responsibilities in the area with the PB4Ys of VB-108 and 109, the Catalinas of VP-72 and 53 and the PBMs from VP-202.

On completion of training on the West Coast in August 1943, VB-140 arrived at Henderson Field, Guadalcanal, and commenced flying patrol and strikes in the area. Early the following month, a six-plane detachment was deployed to the Russell Islands. In addition to routine search missions, the detachment flew frequent strikes, bombing and strafing the numerous active Japanese bases in the Kahili and Kolombangara areas further up the Solomons chain. VB-140 was joined in the Russells by VB-138 in October 1943, with the two units sharing responsibility for keeping pressure on enemy bases in the Solomons.

On 19 December 1943, the first three Venturas of VB-142 took off from Kaneohe Bay, bound for Tarawa. This advance guard arrived at the coral air strip of the former enemy base on 21 December - just one month after one of the bloodiest battles of the Pacific war had taken place and only a few days after the island had finally been declared secure. The Seabees had just completed resurfacing the landing strip and laying Marston matting in aircraft parking areas. VB-142 immediately began flying sector searches for enemy surface craft and submarines, plus carrying out low-level bombing and strafing raids on Japanese bases in the Marshalls. Because the enemy began risking precious cargo vessels in an obvious effort to supply and strengthen their Marshall Island bases in anticipation of invasion the hunting for VB-142 was good. During this period, eleven enemy freighters were sunk or heavily damaged. A number of patrol and picket boats were also sent to the bottom. The first of these was left sinking by LT Dave Walkinshaw and his crew, off Mille, on New Year's Day of 1944. LCDR Worner caught one at Amr Atoll, LT Swenson sank another at Alinglapalap and LT Williams sank one at Jaluit.

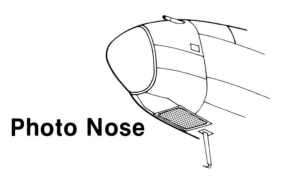
Photo Nose

Tarawa in January 1944, VB-142 and -144 crews lived in tents near the beach to the right of the runway. The single runway was 3800 feet of crushed coral and had to be constantly wet in order to control dust. Note two PBY-5As in upper left. (National Archives)

Co-author LT W.E. 'Bill' Scarborough's Ventura on Tarawa, 1944. Coral dust has scoured the paint from leading edges of the propeller blades. Bugs Bunny was hand-painted by Bill using paints available on Tarawa. (Authors)

VB-144's LT Jim Buie (center) from Los Angeles and his well-armed crew in front of Pistol Packin Mama, **BuNo 34805, on Tarawa 1944. The Tommy gun was carried for possible use in a survival situation if forced down over enemy territory. (Authors)**

A VB-137 aircraft on the runway at Nanomea Is. in the Ellice Group. This early-production PV-1 shows recently-added bars to the star on the nose, probably done in the field after the change was ordered on 28 June 1943. This idyllic setting of waving palms was not typical of forward area bases! (National Archives)

Despite the many flights by lone Venturas through these areas, there was only one major attempt at interception. LT Hureger and his crew were assigned to a reconnaissance search over Taroa to obtain information on enemy strength and disposition. As the PV-1 approached to within a quarter mile of the airfield, five Zeke fighters scrambled to meet it. Hureger was tempted to take them on, but, with only enough fuel on board to get back to his base, some 500 miles away, he reluctantly turned away. In the chase that followed, the enemy fighters were never able to close to firing range and eventually turned back.

By the end of January 1944, the stage was set for the next step in the island-hopping Pacific campaign, the invasion of the Marshalls. VB-142's Venturas supported the operation by strikes against the major enemy air base at Mille, on the 29th and 30th. These raids effectively eliminated the threat of enemy air attacks staging through Mille to attack the US invasion force headed toward Kwajalein.

With the capture of Kwajalein and Majuro, operations settled back to routine for VB-142, punctuated by occasional strikes against Mille, Wotje and Jaluit. In April, the squadron established a detachment at Majuro and initiated a series of raids against Taroa to keep that airfield out of commission. The Japanese base on Nauru posed a similar threat and on 3 May VB-142 launched a sunset strike with all twelve aircraft. This long-bypassed stronghold had not been subjected to the crushing blows delivered to other bases and its AA defense remained the most deadly in the Marshall area. Despite the concentrated AA fire, which bracketed every plane during their bombing runs, none were hit and nearly nine tons of bombs were dropped on the airfield. During June, strikes continued against Nauru and the Ocean Islands and, though many planes were hit by AA, none were lost and the PV's reputation for an ability to absorb punishment continued to grow.

Meanwhile, on 12 January 1944, VB-144 had arrived on Tarawa to provide additional support for the searches and strikes preliminary to the Marshalls invasion. The squadron suffered its first loss at Midway when a Ventura returning from patrol missed the fogged-in island and was forced to ditch when fuel ran out. The plane was abandoned with minimal damage and no injury to the crew, but it floated for less than 30 seconds and the crew escaped with little more than the clothes they were wearing. Fortunately a liferaft floated to the surface after the plane sank which supported the men until they were found and rescued by a PT boat from Midway.

Operations had begun immediately after the squadron's arrival at Tarawa, with sector searches, rotating daily with VB-142. Interdiction strikes were also flown against Nauru in the Gilberts, Wotje, Taroa, Jaluit and Taongi in the Marshalls and Kusaie in the Carolines. Early in April, following the invasion of Kwajalein, a detachment began operations from Roi, at the northern end of the atoll. Planes rotating to the Roi detachment flew via Jaluit, bombing and strafing targets there to insure against their use by enemy aircraft. From Roi, VB-144 flew two daily 300 mile sector searches and a photo-recon mission over Kusaie. Frequent strikes against Japanese bases in the Marshalls were flown, coordinated with the Corsairs of MAG-31, also based on Roi.

The tactic used against land targets was usually a coordinated multi-plane glide bombing run, pushing over at 6000-8000 ft and reaching speeds of 280 to 300 knots (322-345mph) in a 45 degree dive. Bomb loads were usually six 500 pound GP with contact or short-delay fuzing, dropped by the pilot in train. Heavy AA fire was frequently encountered but was not accurate. Attacks on ships at sea or on well-defined pinpoint targets ashore were usually single-plane low-level attacks, with continuous strafing with the bow guns being carried on during the run-in. These attacks produced good results, though at considerable risk to the attacker. Japanese light and medium AA was effective and low-level Venturas were hit frequently. VB-144's only combat fatality occurred during such a raid on Wotje. At 50 feet, with bomb bay doors open, the PV was hit by a burst of .50 cal AA. One projectile penetrated the nose of the aircraft and exploded in the cockpit, killing the pilot instantly. The co-pilot, LTJG Ken McNatt, a replacement pilot on his first combat mission, took control of the Ventura, jettisoned the bombs and flew the plane away from the target area. After rendezvousing with the other strike planes, McNatt flew the damaged plane back to base and made a safe landing - his first ever in a PV!

VB-144's preference for glide bombing versus low-level attacks was shared by the other

Ventura squadrons. USAAF B-25 squadrons operating in the CenPac area quickly learned the same lesson, but at considerable cost. For example, the 396th BS of the 41st BG based on Tarawa in early 1944, lost more than half of its original aircrew killed or wounded in only six missions! Abandoning low-level tactics, the Mitchells moved up to medium altitudes. These attacks were effective against runways and airfield installations, the primary targets of these interdiction strikes.

In mid-July, VB-144 moved to Roi and operated from there until relieved by VB-133 on 1 September 1944. Coordinated attacks with Marine Corsairs against the Marshalls, routine sector searches and a series of propaganda leaflet drops on Nauru occupied the squadron for the remainder of their deployment.

In the SoPac area, VB-140 was relieved by VB-148 at Munda, New Georgia, on 1 April 1944. While operating from Munda, the new squadron flew a variety of missions, flying escort for USAAF C-47s carrying paratroopers into New Guinea and strikes against targets on Bougainville. The C-47 escort missions were flown at irregular intervals and were never opposed. For these, the PVs utilized standard fighter escrot tactics, flying a weave over the C-47s. Flights to Bougainville were to harrass the enemy by bombing and strafing targets of opportunity. The search planes were normally loaded with four 500 pound GP bombs and flew the assigned route at tree-top level at 200 knots. Enemy troops fired small arms at the planes and it was not unusual for them to return to base with multiple 7.7mm holes.

On 14 April 1944, LT W.T. Henderson and his crew took off from Munda on what was to have been a typical search of the Bougainville area. Henderson proceeded on course, investigating several hulks along the shore, then turned toward Kahili, one of the active and heavily defended Japanese air bases. As the PV crossed over the enemy base it was fired on by AA, receiving heavy damage. Unable to continue flying, Henderson turned away from the island and headed out to sea, ditching the Ventura about four miles from Ballale Island, another enemy stronghold. Wounded turret gunner Wood went down with the plane, the rest of the crew made it into the raft. The mechanic, J.R. Elkey, also wounded, died in the raft. The pilots and the radioman were picked up by a Dumbo rescue plane and, although suffering from burns, were able to return to duty after about three weeks.

On 21 May, VB-148 was transferred to Emirau to extend its search toward the Western Carolines. On one of the long range search flights, LT Harry Metke engaged and shot down a Betty bomber south of Truk. A week later, LT Harry Stanford shot down another:

> On 8 June 1944, while on ASW patrol from Emirau, my Ordnanceman sighted an unidentified aircraft several miles distant on an opposite course to ours and at about our altitude of 500 feet. We gave chase and closed low from his rear quarter and identified the aircraft as a Japanese Betty - probably out of Truk. I jettisoned our depth charges and closed to a close range before opening fire with the twin Cal. 50 bow guns. Almost immediately the Betty caught fire at the Starboard engine nacelle. I was closing fast and overshot as the Betty reversed course thereby giving me another opportunity for a firing pass coming in again from his rear quarter. The Betty lost altitude, struck the water, exploded and burned. On the first pass, the Betty returned fire, one bullet entering the starboard side of the cockpit, grazing the copilot's leg and striking the pedestal severing a main electrical circuit there. We also received several other holes in the starboard wing and aft fuselage section.

On 26 July, LT Metke sighted a convoy southwest of Truk while on one of the daily search flights. After radioing a contact report, Metke attacked the convoy with depth charges, the only weapons he had, and succeeded in damaging two of the ships. The radioman, Charles Simpson, was wounded during the attack. Later in the day, a six-plane strike, led by LCDR Jakeman, located the convoy and with low-level attacks sank three ships and damaged a fourth. The following day, another six-plane strike attacked the remnants of the convoy and sank a fourth ship, destroyed a defending Ki-61 Tony and pro-

VB-142's Commanding Officer, LCDR John H. Guthrie, and his crew. Flight clothing for PV crews ranged from heavy winter flight gear in the Aleutians to lightweight flight suits (coveralls), dungarees, Army shirts and hastily improvised shorts (trouser legs cut off) in the tropics. (USN via Warren B. Herrick)

bably destroyed another. The fighter destroyed was shot down by turret gunner Leonard Wheatley. As the Tony turned in toward LT Van Wilber's Ventura, Wheatley tracked him and fired one long burst from the twin .50s. The Tony exploded in a brilliant flash before it had closed to firing range. VB-148 received congratulatory messages from its operational commanders for this outstanding performance - including one from General Douglas MacArthur!! VB-148 was relieved by RNZAF squadron on 15 October 1944, and on 22 October the first planes started the return flight to Hawaii.

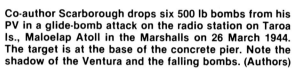

Co-author Scarborough drops six 500 lb bombs from his PV in a glide-bomb attack on the radio station on Taroa Is., Maloelap Atoll in the Marshalls on 26 March 1944. The target is at the base of the concrete pier. Note the shadow of the Ventura and the falling bombs. (Authors)

A VB-142 PV-1 in a typical revetment on Tarawa. There were few revetments on Tarawa due to the lack of space and the time required to build them. Those available were usually reserved for the 'ready' planes. Shelters for standby crews were usually built into the revetment walls, hence the vents visible above the wing of the Ventura. (Warren Herrick)

Blonde Blitz, a late Ventura of an unknown unit in the Central Pacific, carrying the transition-style national insignia (without surround) and the later three-color camouflage. (Bowers)

VB-144 PV-1, BuNo 33374, Lady Luck, enroute to a Marshall Is. target from Tarawa in 1944. (Authors)

A 20mm AA hit on the port engine riddled the fuselage of this PV, which made it back to base on a single-engine. Note feathered prop blade. (Authors)

(Below Right) VB-144's LT Rook flew 34730 with this nose art, entitled Senorita Ventura. (Authors)

Jimmy Junior, a PV-1 flown by LT Jim Brady and his crew from VB-144, sported the usual pretty girl on the plane's nose. The crew wanted the girl and Brady wanted to name the plane for his new son back in the States! A compromise was reached. (Authors)

VB-148's Executive Officer, Harry Stanford, and crew after shooting down a Betty (Mitsubishi G4M2 Naval bomber) south of Truk. Stanford is on the left in the back row. Note the ball mount for a machine gun under the national insignia. (USN via Harry Stanford)

(Below Left) First firing pass - flame errupts from the Betty's starboard engine nacelle. The pictures were taken by the copilot, Thibodeau, who was hit in the leg but continued to snap away. (USN via Harry Stanford)

(Below) Betty, tail number 01-322, just prior to crash. (USN via Harry Stanford)

(Bottom) Betty hits the water, explodes and burns. Stanford circled the burning wreckage but saw no evidence of survivors. (USN via Harry Stanford)

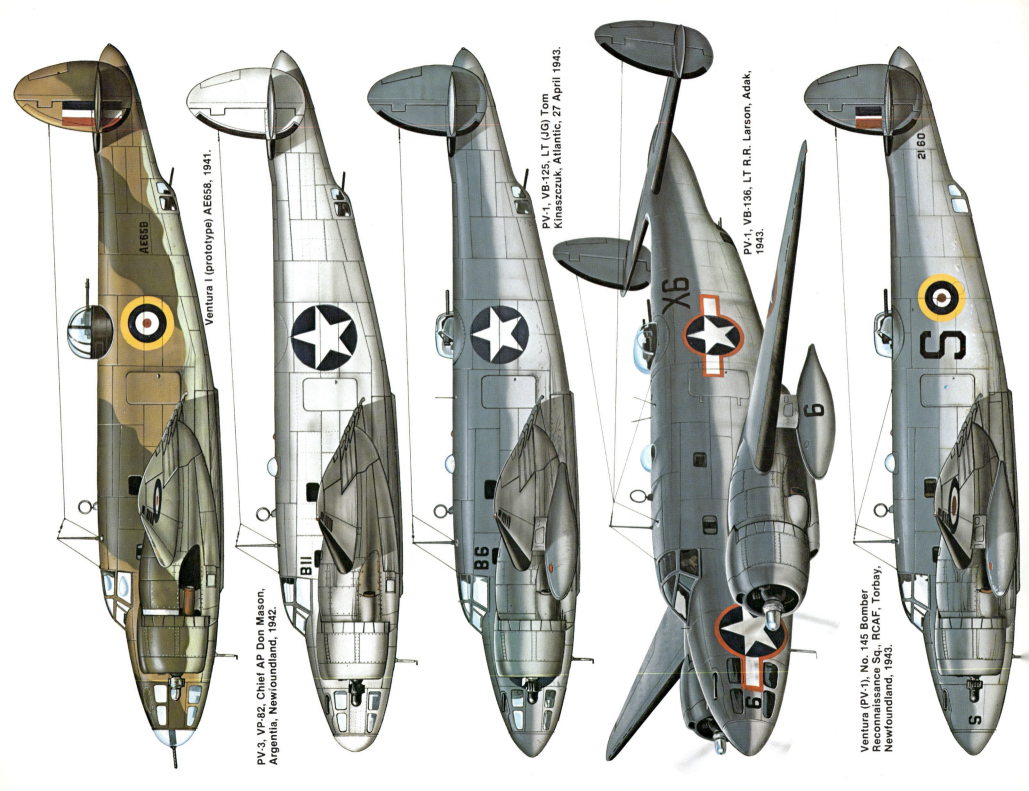

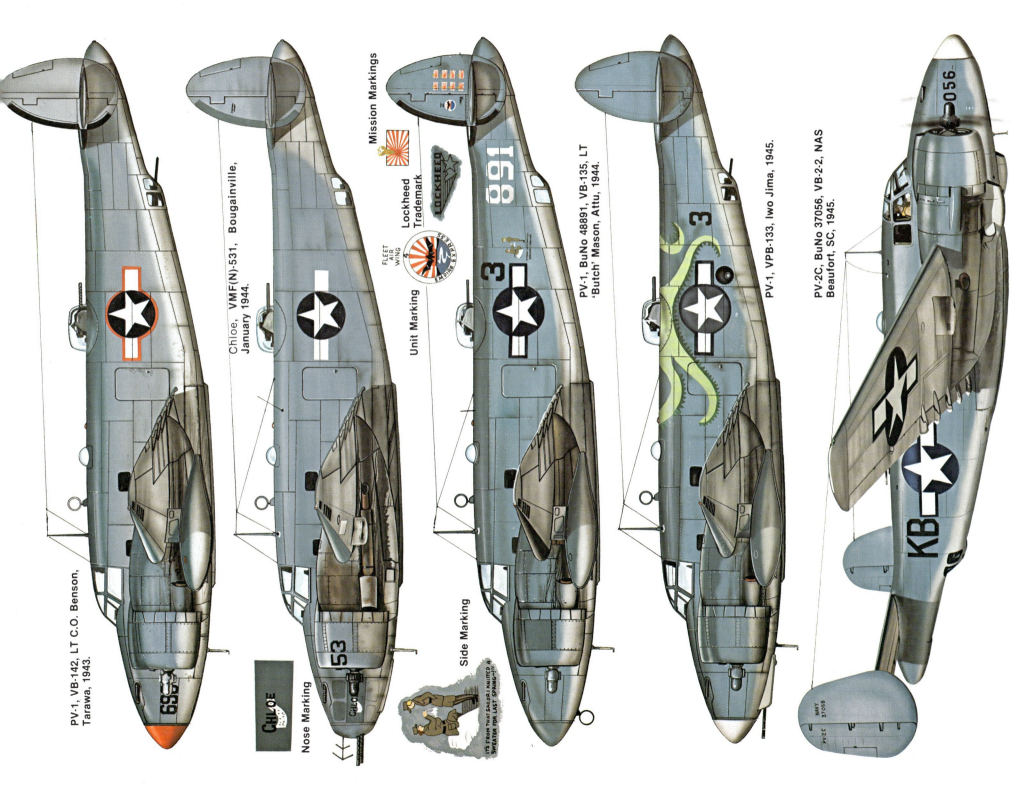

Very early VB-136 Venturas with barless national insignia, Adak, 1943. The color scheme is Blue Gray over Light Gray.

The Aleutians

The first deployment of Venturas in the Pacific was to the bleak, fog-shrouded Aleutian Islands. VB-135 was first, arriving at Adak in April 1943, followed eighteen days later by VB-136. Later, VB-135 moved to Amchitka to fly missions over Japanese-occupied Kiska, while VB-136 remained at Adak. Because of the ever present fog VB-135 Venturas, with their ASD-1 radar, acted as 'Pathfinders' for B-24 Liberators of the 11th Air Force in bombing missions to Kiska. This highly effective method of radar-bombing was developed and perfected by VB-135.

In May 1943 American forces invaded Attu and while the fighting still raged, Army Engineers were building an airfield at Casco Cove on the eastern side of the rugged island. Not until August, (this gives some idea of the difficulty of conditions) had the Engineers finally laid enough Marston strip for a 3000' runway so VB-135 could move in. After a raid by Japanese Bettys based in the Kuriles, VB-136 moved up to Attu as well. The two Ventura squadrons then began flying anti-aircraft patrols toward the Kuriles to guard against further air raids. On two occasions Bettys were attacked and turned back by the Venturas.

VB-135 and -136 were relieved by VB-139 in December 1943. Early in January 1944 LTs R.A. McGregor, T.R. McKelvey and D.M. Birsall of VB-139 began pioneering night photo and bombing missions to the Kuriles. These nightly flights were made under the most hazardous conditions imaginable - arctic winter weather with snow, sleet, fog and icing conditions, ice-covered, fogged-in runways and 9-to-10 hour flights over the trackless North Pacific and it was sure death in the freezing waters if forced down. Nevertheless, VB-139 successfully completed 78 photographic and bombing missions over the Kuriles before being relieved in May by VB-135 now back at Attu for a second tour of duty. 'Bombing 135' continued the nightly missions to the Kuriles until June 1944, when a daylight bombing raid to Shimushu demonstrated that daylight missions could be more effective. In June 1944, VB-135 was joined at Attu by VB-136, also back for a second tour of duty. VB-136 was now mounted on an improved PV-1 with more sophisticated radio equipment, an automatic fuel system and three additional 'chin' guns under the nose. Regular daylight strikes by the two squadrons were carried out with shipping as the primary target, but if none was to be found, land targets on Paramushire and Shimushu were pounded. Japanese fear of an American invasion through the Kuriles mounted, and fighter aircraft were moved to the Kuriles to combat the Venturas.

The Venturas of VB-135 and -136 weren't deterred, regularly outrunning, and frequently shooting down, the enemy fighters. At no time did the PVs have fighter escort. They were ordered to attack the Japanese fighters only in self-defense. At least those were the orders. LT Pat Patteson of VB-135 on solo daylight photo missions over the Kuriles on 14 and 25 June 1944, shot down a Hamp with his bow guns while his turret gunner, AOM2c Floyd Jacobsen, bagged three more! LT J.W. Pool, also of VB-135, was attacked by eight Tojos over Shimushu. He shot one down in flames and outran the remainder. VB-136's LT F.R. Littleton shot down a Tojo in a head-on attack on two of the fighters. After these encounters, the enemy fighter pilots became less aggressive and seemed reluctant to tangle with the Venturas. They still chased them...but usually just out of gun range. VPB-131 relieved VB-135 at Attu in October 1944. The new squadron was equipped with fifteen of the last, and best, PV-1s. They also brought a new weapon to the Aleutians, the 5" HVAR rocket. VPB-131 operated the Ventura as an attack plane, carrying no bombs. The bomb bay was completely filled with 480 gallons of additional fuel. Their only weapons were the 5" rockets and .50 cal mgs. USAAF resumed bombing the Kuriles, recruiting VPB-131 to fly diversionary escort missions for the B-24s. The strategy was for the Venturas to strike first at another location to draw the enemy fighters away from the trailing, high-altitude B-24s. These tactics were successful for several missions. After the beginning of January 1945, VPB-131 Venturas flew four-plane rocket strikes to the distant Kuriles without the Liberators.

These successes were not without cost. VPB-135 lost more planes and crews than any of the other Aleutian PV squadrons - ten planes. However, some of them diverted to Russian Siberia, the only alternate to flying back across the North Pacific to Attu. Their treatment as virtual POWs by our Russian Allies is a harrowing story. The crews were eventually returned via Europe. The Aleutian weather, however, constituted a much greater threat than the Japanese. In over two years of battling arctic cold, snow, ice, fog and Willi-Waws, as well as the Japanese, the PV squadrons lost more men and airplanes to the weather than to the enemy.

Summer 1943 on Adak. A lone PBY-5A Catalina is surrounded by VB-136 Venturas. The PVs are still in 1942 paint scheme of Blue Gray over Light Gray with national insignia having been updated in the field. (USN via Robert L. Lawson)

VB-136's No. 3 in flight. Note how the added bar to the national insignia covers the side window. Also of interest is the Donald Duck cartoon on the rear of the fuselage and the aircraft number being repeated on the drop tank.

A VB-139 photo-Ventura takes off from Attu for a mission to the Kuriles in January 1944. Due to its speed, the Navy made extensive use of the PV-1 as a photo-reconnaissance plane. No PV-1P designation was officially assigned since all were field modifications. A large K19A camera was mounted behind the flat glass panel under the nose. Camera hatches were provided in both PV-1 and PV-2 fuselages and as many as three large F-56 cameras could be installed for vertical and oblique photographs. (National Archives)

VB-135 crew donning heavy winter flight gear for a mission to the Kuriles. Aircrews were issued S&W 38 cal. revolvers. Lt. 'Butch' Mason's PV-1 is in the background. (USN via L.A. Patteson)

A pair of VB-135 Venturas enroute to Paramushiro from Attu in July 1944. The white '936' is the last three digits of the BuNo (48936); black '10' is the plane-in-squadron number. The loop antenna under the nose was installed to combat precipitation static which was a serious problem in the Aleutian area. (USN via M.A. 'Butch' Mason)

VB-135 Venturas in a three-plane revetment on Attu. It's Summer 1944 - the snow is nearly gone - only the fog remains to make flying treacherous. Note that only aircraft No. 8 has the plane-in-squadron number repeated on its nose. (National Archives)

Turret gunner (Aviation Ordnanceman) cleaning the plexiglass of his turret prior to flight. The turret was the only good location for lookouts, other than the cockpit and on patrols this position was always manned. (Robert R. Larson)

A gasoline heater is used to thaw out a VB-136 Ventura prior to a mission to the Kuriles. (National Archives)

VPB-131 PV-1 over the Kuriles. VPB-131 had some of the last, and best, PV-1s manufactured. This one, BuNo 49654, was the fifth from last PV-1 made and the PV-2 was already in production. (A.A. Hoffman)

Night Fighter

In the night skies over Europe a deadly cat-and-mouse game of electronic warfare was waged between the RAF and Luftwaffe. On the basis of this experience, both US Army and Navy embarked on separate night fighter development programs. Progress in the Pacific was slow until the Japanese began their nightly 'Washing Machine Charlie' nuisance raids on Guadalcanal's Henderson Field. Night fighter priority suddenly skyrocketed. The Marine Corps, feeling an immediate need, took the lead in the Navy's program by organizing its first night squadron. VMF(N)-531 was commissioned at MCAS Cherry Point, NC, on 1 April 1943. The Marine Corps surveyed the aircraft available to them which might make suitable night fighters and chose the PV-1 Ventura. The PV-1 possessed the speed, maneuverability and firepower required, its primary disadvantage being a low service ceiling. It was anticipated that most night interceptions would be made above 25,000 feet. In actual combat, Japanese obliged by staying well below the PV-1's ceiling. The Ventura's inability to slow down quickly proved to be its major liability as a night fighter. VMF(N)-531 operated the Venturas with a three-man crew - one pilot, a combined radio-radar operator and a turret gunner.

Night fighting was a difficult, team-effort business. A complex, new system was developed involving not only the crew of the night fighter but also ground radar and a controller who vectored the fighters to the 'bogeys'. Successful night kills were equally dependent upon the skill of the ground controller as the skill of the night fighter crew.

On 15 May 1943, VMF(N)-531 departed Cherry Point for San Diego, its training far from complete and its aircraft not ready. But its presence in the combat zone was urgently required. By 18 October, the squadron, its ground crew and ground radar, was installed at

A VMF(N)-531 fighter on Bougainville in January 1944. **Chloe** shows the external modifications to an early PV-1 to equip it for the nightfighter role. The additional guns and the AI radar antenna in the modified nose cone are the more obvious changes. Note also the blacked-out nose windows and the shield just inboard of the under-wing landing light. Chloe's **markings and camouflage are unusual.** She still carries a worn Blue-Gray & Light Gray scheme. Note also the fuselage star and bar without surround. Both the early war camouflage and unusual treatment of the national insignia were typical of USMC aircraft at this time. (National Archives)

ASV 'Yagi' Antenna

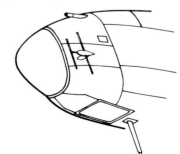

AI IV Radar

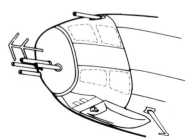

Modified Nose with Four Additional .50 cal MGs

Vella Lavella, in the Solomons. Although VMF(N)-531 was the first night fighter squadron deployed in the Pacific, it was not in the cards for them to score the first American night kill. LCDR Gus Widhelm's Navy F4U squadron, VF(N)-75, scored first on 31 October 1943. However, it was VMF(N)-531's ground controller who vectored the F4U-2 to the kill. Two weeks later, CAPT Jenkins scored the first night kill for VMF(N)-531 in a PV-1. One of the early lessons learned was that night fighters must get dangerously close to their victims before opening fire in order to insure positive identification and a certain kill. The PV-1s frequently returned covered with oil and debris from their victims.

Exerpts from ComAirPac AIR OPERATIONS MEMORANDUM No. 25

PV vs RUFE

During the next ten minutes, the GCI had no bogey info' and sent me to...Motupena Pt., where I was presently vectored on another bogey. I made AI contact at 5000 ft, but lost it as the bogey turned. When our plots separated on GCI, we were on opposite courses. I was again vectored astern of the bogey and made AI contact at one mile. This time I had slowed to 155k, and the target appeared to be well above me. I climbed 500 ft, closing slowly and following the plane through constant changes of heading of about 30 degrees. When I made visual contact, I discovered that the Rufe was making gentle 'S' turns...look(ing) for shipping. I eased in behind him...closed to about 20 yards and opened fire, setting the fuel tank...afire. The plane dove and crashed into the water, still burning.

SIXTEEN ROUNDS DESTORY JAKE

...AI contact was made at 6000 ft. Col. Schwable came in from stbd to port and below. He made a visual at 2000 ft with the bogey about 1000 ft above. At first it looked like a bright star moving slowly across the sky. Col. Schwable had Sgt. Ward vector him...visually while he put the PV through an easy 'S' turn to lose speed. He pulled up and behind the bogey and when within 700 ft the PV was almost at stalling speed. Sgt. Ward had now returned to his scopes and when he read 700 ft, he told the pilot to 'let him have it', and then again left the scopes to watch Schwable close to within 300 ft before touching-off his lower guns. Col. Schwable fired the exceptionally short burst of eight rounds and was followed immediately by Sgt. Fletcher's two turret guns with an eight round burst. Instantly, the bogey flamed and the wings flew back and up...and the two meatballs were plainly visible on the top of the wings. The Jake fell off, as the PV pulled up and to the right, and flaming like a torch, dropped off into the water.

ONE LESS BETTY

...an AI contact was made at about 12000 ft range. The bogey appeared at this point as one large blip and Sgt. Kinne notified the pilot that the bogey was a large target. LCol. Harshberger closed in...as directed by Kinne. Within eight minutes the PV had closed to within 4000 ft of the bogey which at this point developed on the scope to be two blips. A few moments later at 2500 ft, Harshberger and turret gunner Tiedeman made a visual of two Betty-type planes flying formation. Harshberger closed to 1500 to 2000 ft when the Betty on the left started in with its tail gun. LCol. Harshberger picked the Betty to his left and closed to 1000 ft. He opened up with his six nose guns and fired one burst into the belly of this bogey. Sgt. Tiedeman also put one burst into the tail of this Betty and concluding that the pilot had this one in hand swung his sight onto the one on the right. At this point the PV was hit, in the nose, putting five of the six guns out of commission. Harshberger continued to fire with the remaining gun. Tiedeman put one burst into the tail gunner of the Betty on the right and it peeled off to the right and began to get out of range. He then swung back on the bogey to the left and estimated it to be 500 ft away and 15 degrees up. He put three bursts into the bogey...The relative motion of both planes was very slow at this point and the bogey was slowly maneuvering up and to stbd. Sgt. Tiedeman...put one last burst into the Betty. The Betty had started to glow internally. The intensity of the glow rapidly increased and the fuselage of the Betty took on the appearance of a brightly lighted sieve. ...Harshberger dropped back to watch the descent of the bogey which by now was a flaming mass.

When VMF(N)-531 was relieved in June 1944, the squadron score stood at twelve confirmed night kills. They had been the pioneers in a complex, new form of warfare, their success a testimonial to their dedication and ingenuity in adapting the Ventura to a role never contemplated by the aircraft's designers.

A pair of VMF(N) PV-1s at Bougainville late in 1944. The Venturas being used by the Marines by this time closely resembled standard machines. (National Archives)

A four-plane strike of VB-128's PV-1s enroute to Brunei, Borneo, 24 May 1945. These Venturas have plane-in-squadron numbers on their nose and the last three digits of their BuNos on the tail.

The Philippines

By October 1944, American forces were advancing toward the Philippines in a two-pronged offensive. In support of the invasion of the Philippines, VPB-137, back for a second tour of duty in fifteen new PV-1 Venturas, departed Kaneohe Bay, Hawaii in October for Los Negres where they began flying daily searches of the area. On 29 November Japanese planes strafed Momote and a special search was flown by VPB-137 toward Rabaul and Kaveing for enemy aircraft, but none were found on any of the airstrips in the area. That same day the squadron moved to Morotai in order to fly more effective searches. On 6 and 7 December, special searches were flown for a PBY missing along the coast of Mindanao, but no trace was found of the Catalina.

From 3 to 10 December 1944, a series of four plane strikes hit a radio station on Tobi and a suspected aircraft assembly plant at Lalos where Petes had been reported operating. The targets, buildings and repair shops, were hit by glide bombing and strafing runs. Strike photos showed both targets destroyed and the Army reported that the radio station ceased broadcasting with the first strike.

On 14 December, LT Enevold spotted a Pete taking off near Bongao. He firewalled the throttles and from five miles away closed on the Pete as it was making a climbing turn. He fired a 30 degree deflection shot with the five bow guns and the Pete began a spiraling turn back to the water. Enevold pulled around and made another firing run, leaving the Pete blazing and sinking.

On 19 December, near Bongao, one of those regrettable wartime incidents occurred, the only known combat encounter between P-47 Thunderbolts and Venturas. While cruising along on the outbound leg of their sector, LT Hancock's PV-1 was jumped by three P-47s. The Ventura crew had been drilled in aircraft recognition and recognized the Thunderbolts while they were mere dots in the distant sky. They watched with amusement as the P-47s closed and began setting up a firing pass on them. The PV could have given the P-47s a run for their money, but, thinking that they were merely horsing around, Hancock did not increase speed. This being a war patrol, though, the turret was manned and the guns charged. All three P-47s continued to close on the PV-1 but two broke off, apparently recognizing the Ventura as a friendly or seeing the 48 inch diameter national insignia. One, however, continued his run and opened fire. The PV-1 turret gunner returned fire on the attacking P-47, firing two bursts, shattering the canopy and wounding the Army pilot. The P-47s joined-up and escorted their wounded 'Ace' back to their base where he was able to land. The Ventura sustained little damage, excepting a shot-out tire, and returned to Morotai. Due to the flat tire, the port landing gear failed on landing, the PV ground-looped and was totally destroyed. Fortunately the crew was uninjured and all six walked away from the crash. It is not known if any official action was taken.

The New Year of 1945 was ushered in by two events, VPB-128's arrival at Owi to reinforce VPB-137 and a night bombing raid by the Japanese. Nine of VPB-137's PV-1s were destroyed in the raid, and two more were later written off, eliminating eleven of the squadron's fourteen planes. On 5 January, twelve of VPB-128's Venturas were flown to Tacloban and transferred to VPB-137, leaving the former squadron with only one aircraft.

With their new Venturas, VPB-137 resumed flying daily searches, including a dawn and dusk anti-sub patrol, from Tacloban, on Leyte. LCDR Porter, flying the dusk anti-sub patrol, located two freighters off Dumaguete. He made strafing and rocket runs on both, scoring direct rocket hits on one. During another strafing run on the other ship, a large fire erupted from the hold. Two hours later, on the return leg, the fire-gutted hulks were seen still burning.

LT Stanley Miller did what Lockheed engineers said was impossible in a PV-1. During take-off, shortly after becoming airborne, one of his engines failed completely. In the excitement of keeping the PV-1 in the air and right-side-up, Miller and his crew failed to jettison bombs or drop their auxilliary tanks. Design studies had indicated that a PV-1 could not fly on one engine with such a load, but Miller circled the strip at a very low altitude and settled back on the runway with no damage except to the crew's nerves.

The most unorthodox Ventura ditching occurred on 24 February 1945, when LT Enevold and crew of VPB-137 located a camouflaged Pete floatplane moored close to an island in the Philippines. During a strafing run on the Pete, two of the chin guns jammed. Distracted, Enevold continued flying at low altitude down the coast. The crew ordnanceman (AOM) squirmed down into the nose to clear the jammed guns. Gaining access to the nose compartment in flight involved sliding under the copilot's rudder pedals. Enevold became so engrossed in watching the AOM that he failed to monitor his altitude and flew into the water while doing 160 knots. The impact ripped off the bomb bay doors but bounced the PV high enough that the pilot was able to make a normal landing on the water without flaps. In the few seconds between the first impact and the second, all members of the crew, including the AOM in the nose, were able to reach their ditching stations. The AOM must have left footprints on the copilot's face as he literally exploded up out of the nose! The plane sank within 20 seconds after coming to a stop on the water. Native Philippine fishermen, who witnessed the ditching, rescued the crew in a matter of minutes. The following morning a PBY landed at the fishing village, picked up the crew and returned them to Tacloban.

Beginning in March 1945, VPB-137's operations were divided between Tacloban, Morotai and Clark Field. Four daily searches were flown out of Tacloban, one out of Morotai and three out of Clark Field. On 6 April, LTJG Locker attacked a pair of Oscars near Amoy. He shot one down in flames and damaged the other which escaped into nearby clouds. LT McAlhany sighted an Oscar from a distance of five miles. He closed to within two miles, but lost him when the enemy pilot ducked behind some hills.

The Tacloban contingent moved to Samar about 15 April flying searches and offensive strikes to the China coast. LT Keach bombed a large vessel in the mouth of the Canton River. LT Enevold located a freighter in a bay on the China coast. He fired his rockets in salvo scoring direct hits. LT Deiss made strafing and rocket runs on a 400 ton vessel near Swabu. Flying debris from his rocket strikes damaged the starboard engine and forced a return to base. LT Deiss failed to return from a search of the Amoy sector. The following day an all-out search effort was mounted. LT Markham located a dye marker and what appeared to be three survivors of a plane crash. He directed a rescue submarine to the scene and flew cover until the survivors had been taken aboard. They were not LT Deiss' crew, however, but the crew of an Army B-25 which had been shot down by enemy fighters.

By May 1945, VPB-137's strength was down to seventeen crews and twelve PV-1s. Operations continued out of Samar and Clark Field, with six planes at each base. On 10 May, patrols out of Clark Field were cancelled and VPB-137 flew a series of strikes against Formosa and northern Luzon. On 27 May the Clark Field contingent joined the remainder of the squadron at Samar. As a final contribution to the war effort, VPB-137 flew rocket and bombing strikes against targets on Formosa until 8 June, when the entire squadron departed the war zone for Hawaii.

VPB-128 whose wings had been clipped by the transfer of all but one of their planes to VPB-137 in January 1945. It was not until their arrival at Samar in March that they received enough aircraft to begin operations. On 18 March, LTs Dorrington and Snyder attacked a pair of submarines in Davao Gulf, scoring one kill and one probable. On the 21st, three VPB-128 Venturas attacked another sub at Cebu, damaging it with rockets. The next day LCDR Tepuni led a follow-up five plane bombing strike. Tepuni made the first run and scored a hit but was shot down by shore AA. As a VP-82 Ensign flying a Hudson, Tepuni had made the first successful attack by US naval aircraft on a German submarine, he and his crew sinking U-656 off Newfoundland on 1 March 1942. In an ironic twist of fate, after surviving three years of combat flying, Tepuni was killed in another attack on a submarine just months before the end of hostilities. After Tepuni's crash LT George Hall made a run and destroyed the sub with three direct hits.

A combination bombing, napalm, rocket and strafing attack by VB-128 PV-1s on an industrial plant at Brunei, Borneo, 24 May 1945. The 'bat-wing' shape and Fowler flap tracks identify the attacking plane as a Ventura. The attack resulted in the complete destruction of the factory. (National Archives)

A Japanese freighter under strafing attack from the bow guns of a PV-1. Note debris flying from the bridge. (USN)

On 5 April 1945, VPB-128 moved to Puerta Princessa, Palawan, and commenced strike missions against Japanese shipping. After 28 April, the squadron flew pre-invasion strikes against land targets on Japanese-occupied Borneo, working under the operational control of the 13th Fighter Command, Army Air Forces. These strikes were usually composed of from five to seven PV-1s and an equal number of Army P-38s. The Venturas were armed with two 1000 lb. napalm bombs, eight 5" HVAR rockets and three 500 pound GP bombs. On 21 June 1945, VPB-128 was transferred to Tinian where they spent the remaining month and a half of the war on search missions.

VPB-150 dubbed themselves 'The Devilfish PViators' and painted all their Venturas in a distinctive octopus paint scheme. The devilfish had green tentacles painted on the top of the rear fuselage utilizing the turret dome for the head and the machine gun vents for the eyes. The green tentacles draped over the fuselage in all directions. VPB-150 left four of their PV-1s on Tinian where they were adopted by VPB-133. This one is "Hewego". (USN via Jack Coley)

VPB-133 crew posing before Sea Deuce, **an ex-VPB-150 PV-1 on Iwo Jima. Front row: Warnlof, Moody, Hammett, Gregg. Back row: Duke & Fischer. Note that Moody, the Ordnanceman, is carrying a M-1 Carbine for a survival weapon. These Venturas have the added waist gun positions seen earlier on VB-148 aircraft.** (USN via Jack Coley)

The Marianas and Beyond

The high point of PV operations in the Pacific came with their deployment to the Marianas on the far western edge of the vast Pacific theater. In July 1944, VB-150 relieved VB-142 on Tarawa and VB-133 relieved VB-144 on Roi in August. VB-151 joined VB-150 on Tarawa in August and the three squadrons continued strikes against Japanese bases in the Marshalls and Gilberts, and on Nauru and Wake Islands. In August 150 moved to Tinian in the Marianas shortly after the area had been declared secure and at the height of the rainy season. On the 30th VB-151 arrived and joined VB-150 at North Field, a former Japanese air base. As a safeguard against enemy attacks on the newly captured bases on Tinian and Saipan, the Navy PV squadrons flew daily searches and strikes against the Japanese airfields on Pagan, Yap and Woleai.

VB-133 continued search and strike missions in the Marshalls for the rest of 1944 and into early 1945. Strikes against Wake continued until March when the squadron also transferred to Tinian, to relieve VPB-150. After settling in at their new base, VPB-133 flew searches to the west and southwest of the Marianas and carried out strikes against Woleai, Puluwat and Lamotrek.

Early in 1945, as B-29 attacks from the Marianas increased, the Japanese countered by stationing a fleet of fast, armed picket boats in the waters south and east of the Home Islands. These picket boats were out beyond radar range and provided early warning of incoming raids. Since their presence was a threat to the success of the B-29 missions the two PV squadrons at Tinian were assigned the task of providing anti-picket sweeps. On 14 March, a pair of Venturas from VPB-151 inaugurated the new mission with attacks on two pickets east of Iwo Jima. Rocket and strafing attacks sank one and left the other boat dead in the water.

On 23 March, a detachment of six VPB-133 Venturas was deployed to newly-captured Iwo Jima to continue coordinated PB4Y/PV attacks against the picket boats. On the first of these attacks, LTJG Wilson engaged two of the little boats, sinking one severely damaging the other. During the attack, Wilson and his copilot, ENS McCarthy, were both seriously injured by AA fire. The pilots and their crew were saved by the prompt and courageous action of the enlisted aircrew who rendered first aid and one of them, H.M. Sadler AMMlc, succeeded in flying the Ventura back to a safe landing at Iwo. VPB-133's Skipper, LCDR Christman, with the detachment at Iwo, was killed in an unfortunate non-flying accident when a landing P-51 went out of control and crashed into a group of officers standing near the runway. The Iwo group was recalled to Tinian and, for the remainder of the month of March, flew patrols and strikes against Truk and Woleai. On 22 April, LCDR Flannery assumed command. A crew rotation program had been started and by May most of the original 'plankowner' crews had been replaced. Operations continued

Starboard Waist Position

Port Waist Position

(Field Modification)

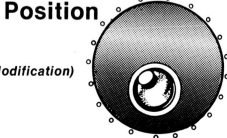

with new faces, old planes and more strikes on Truk and Woleai. In May a nine plane detachment went back to Iwo Jima to renew the picket boat battle. Encounters were frequent, with many of the boats sunk or left dead in the water. In addition to picket boat missions, VPB-133 found itself saddled with daily sector searches which extended from Iwo all the way to Kyushu and Honshu. The squadron flew four searches each day and found plenty of action, particularly in the sectors which included the Japanese Home Islands. On 9 May, LTJG Ray Phillips scored a first when he and his crew bombed and strafed a lighthouse and buildings at Shiono Misake, the first attack by a PV on Honshu. Next day LT Ralph Allen got a freighter in the same area using his HVAR rockets. Defying the 13th 'jinx', LCDR Jack Coley flew the hot sector and attacked two freighters, then flew inland and strafed and rocketed two trains, destroying both locomotives. On the 21st LT Wooten destroyed a bridge and sank a freighter but, as the result of flak damage, ran short of fuel and ditched thirty miles north of Iwo alongside the destroyer USS Cummings. LTs Schenk and Haliburton had the sector on 22 May but became separated enroute. Schenk hit a freighter and left it burning. He was jumped by two fighters, but outran them! Haliburton attacked a radar site on O Shima. The next day, LTs Walker and Burger attacked four small freighters, a pair of Petes and a radio installation near Tanabe. AA fire holed Walker's plane 21 times and one of his crew was injured, but they returned safely to Iwo. LTs Schenk and Duffy paired up for the hot sector on the 27th but became separated over the target area. LT Duffey returned, but Schenk and his crew disappeared. Next day, a massive search effort was mounted. Not only was the search fruitless, but two more planes and one crew were lost. LTJG Phillips had sent a distress message and later was in radio contact with a PB4Y but was not found and not heard from again. Jack Coley on the same search attacked a convoy of eight vessels. His plane was hit by AA fire, forcing him to secure one engine and a hole in a fuel cell reduced range. Coley was forced to ditch about 300 miles from Iwo, but fortunately he had reached a lifeguard submarine. He and his crew were rescued uninjured. Intensive searches for the next several days, with the assistance of PB4Y-2s from VPB-102 and PBY-5As from VPB-23, were unsuccessful and no trace was found of the missing crews or their planes.

Early in June, the Iwo sector plan was revised. Under the new plan VPB-133 patroled mostly west and northwest of Iwo Jima, affording fewer contacts with enemy shipping. Later in June the Iwo detachment was recalled to Tinian. For the rest of the month, and throughout July, VPB-133 flew 'whitecap' patrols enlivened by strikes to Alot, Puluwat, Woleai and Lamotrak in the Carolines. On Puluwat stood a redoubtable target, seemingly impervious to the most determined attack, a lighthouse which had withstood everything from 5'' rockets to 500 pound bombs and had come to be known as 'the poor man's Tokyo'. In mid-July, a detachment of four crews and three PV-1s was sent to Peleliu to fly day searches and night ASW patrols. The night flights included a nuisance raid on Babelthuap, dropping a 100 lb bomb to keep the enemy awake.

In August, a novel weapon was assigned to VPB-133, a PV-1 equipped with a loudspeaker and amplifying system. Flown by LCDR Flannery and a special crew, it was used in flights over enemy territory for propaganda broadcasts. On 24 August, a VPB-133 crew headed by LT Walker was transferred to the senior command in the area, to operate the 'Polly Plane' in flights over the Marianas and the Bonins.

VPB-152 was also based at Peleiu with PV-1s. This squadron found itself involved in one of the most dramatic rescue efforts of the Pacific war when LTJG William C. Gwinn, flying a routine patrol, sighted an oil slick and decided to investigate. Circling the area for a closer look, Gwinn found 30 men in the water, many without lifejackets. He had found survivors of the **USS Indianapolis**. Shortly after midnight on 30 July 1945, the heavy cruiser, having just delivered the first atomic bomb to Tinian, was torpedoed and sunk by the Japanese submarine I-58 in the Philippine Sea. Three days later, on 2 August, LT Gwinn found the survivors. Gwinn's crew dropped their raft and all six of their life jackets to the men in the water, then sent a contact report while continuing to maintain visual contact with the survivors. Gwinn's report initiated rescue operations, and, within a few hours, two Catalinas had landed and begun picking swimmers from the water. A massive rescue effort by surface craft, supported by Navy and Air Force planes, was initiated. By the next afternoon 316 men of the original 1200 man crew, including the Skipper, Captain McVay, had been rescued.

A VPB-152 PV-1 at Shanghai in 1945, just weeks after VJ Day. Note that guns have been removed from 'Chin Gun' package, but the nose guns are still carried. Note also the zero length rocket rails beneath the near wing. (Bowers)

A true rarity, a PV-1 in overall Sea Blue paint! This photo was taken at Shanghai in 1945, which would indicate that repainting was performed on one of the islands in the Marianas. Repair work was the only painting done in the combat zones — and sometimes not even that. Navy aircraft usually wore their factory-applied paint for their entire combat career. Only national insignia and unit markings were kept current. (Bowers)

PV-1 Ventura (USN production run) in Free-French markings.

Dazzle-striped RCAF Ventura target-tug at Trenton, Ontario. Paint scheme is Yellow with Black diagonal stripes. Venturas remained in RCAF service until 1957.

The first Harpoon to roll out of the Lockheed plant at Burbank, CA on 8 November 1943. Two of the main recognition features of the Harpoon, a revised and enlarged tail group and bulged bomb bay are shown to advantage. (Lockheed)

PV-2 Harpoon

As soon as production of the Navy version of the Ventura began at Burbank, Jack Wassal, Lockheed-Vega's Chief Engineer, began design work on a version which would better suit the Navy's needs. Wingspan was increased to 75 ft which substantially reduced the wing loading. Over a ton of payload was added, and fuel capacity was increased with a consequent gain in range from 1650 to 1800 miles. A completely redesigned tail group resulted in a marked improvement in both ground handling and single-engine control, two of the Ventura's weakest points. Wassal claimed a sacrifice of only 6mph in top speed using the same P&W R-2800 engines. However, maximum speed was actually reduced by about 20-30mph compared to the PV-1. Chin guns were standardized on the Harpoon, providing five fixed forward-firing .50 cal mgs, plus two in the dorsal turret and another pair in the tunnel. In order to accomodate the increased payload, the bomb bay doors were redesigned with a noticeable bulge. The greater capacity that resulted allowed a larger load to be stowed internally. Two Tiny Tim rockets could be carried internally by the Harpoon, compared to one for the Ventura. Although slower and less maneuverable than the PV-1, the improved handling qualities — especially single-engine performance — made the PV-2 a superior aircraft for routine operations.

On 30 June 1943, a contract was signed with Lockheed for 500 Harpoons. The first flight of the prototype was in November 1943 with the Burbank plant beginning production of PV-2s in March 1944. PV-1 production continued at Burbank until May 1944. Problems were encountered in sealing the internal wing fuel tanks on the Harpoon's new wing. As a result, the first 30 Harpoons were redesignated PV-2C and assigned to training squadrons with the outboard fuel tanks sealed off. Because of the similarity between the PV-1 and the PV-2, crews transitioning from the Ventura to the Harpoon experienced little difficulty. The improved handling qualities, added comfort and load-carrying ability of the PV-2 were appreciated. However, the loss of speed and maneuverability due to the greater wingspan and increased weight was disconcerting. 'Old hands' could remember how the PV-1 could run away and leave enemy fighters behind, something the Harpoon couldn't quite do.

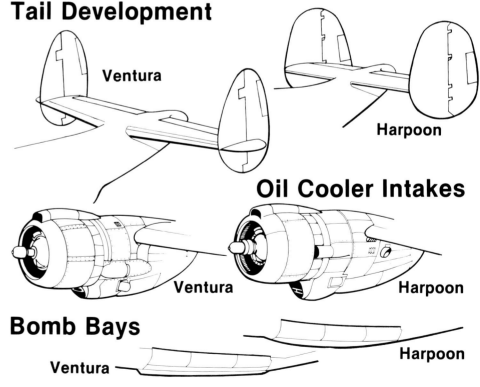

An unusual experiment dubbed PV-1½! A PV-2 tail assembly was grafted onto a PV-1, BuNo 34986. The resulting hybrid offered improved directional control, but required excessive trim changes with power and speed changes. No other examples were produced. (USN via Paul R. Matt)

The PV-2 was rushed into combat before service testing was complete. Fleet operations, however, disclosed a serious flaw in the wing spars. The problem was immediately corrected on the production line, but more than 100 aircraft were in service by that time. The Navy ordered flight restrictions on the Harpoons, but squadron operations were not seriously affected. Dive angle and maximum speed were officially restricted, but these performance limts were seldom adhered to by flight crews. VPB-139 in the Aleutians and VPB-142 on Tinian both had received Harpoons by this time, and both continued combat operations with no ill effects. The early PV-2s and -2Cs with the 'soft' wings were modified by a beef-up consisting of a reinforcing plate riveted over the lower flange of the spars. Some aircraft were returned to the factory for the modification but many had the change made in the field.

One hundred Harpoons were ordered with eight guns mounted in the nose under the designation PV-2D. Due to the war's end and cancellation of contracts only 35 of the upgunned PV-2Ds were delivered.

In March 1945, VPB-139 returned to Attu in PV-2 Harpoons, relieving VPB-136. Harpoons assumed the role of the attacker. The PV-2s carried out both rocket and bombing attacks to the Kuriles, continuing until the end of the war. On 22 June 1945, near Paramushiro VPB-139's LT Marlin scored one of the most unusual kills of WWII. He trapped a Hamp fighter under his Harpoon. Marlin scored first with the bow guns, then he forced the Hamp right down on the water and pinned him there. He rocked the Harpoon over so the turret gunner could get a shot, but he could only stay briefly in that position and keep the Hamp trapped. He then pulled ahead until the tunnel gunner could bring his twin .50s to bear, who delivered the "coup de grace".

VPB-142 returned to the central Pacific for a second tour early in 1945 flying PV-2 Harpoons. During March and April a detachment which had been flying patrols from Midway to gain experience in the PV-2 returned to Kaneohe and the entire squadron departed for Tinian, in the Marianas. VPB-142 started flying patrol sectors and a daily armed reconnaissance of the Truk Islands, beginning in June. Flying patrol is not a fascinating job. After you have seen one square mile of ocean, you have seen them all. The sectors assigned covered the Navy's shipping lanes to Guam and Saipan and on a good day a crew might see fifteen or twenty US ships in the sector. The rest of the time, it was clouds, sky and water as far as the eye could see. On 27 June, LTJG R.C. Janes' crew was crossing the end of their sector and about to turn toward home on the inbound leg, when passing over a rain cloud they saw a surfaced Japanese submarine lying motionless below them. All the luck was theirs. The rain cloud which had hidden the sub from radar detection had apparently hidden the Harpoon from the sub's lookouts. Janes pushed over into a diving turn and until they were well into the attack, the submarine seemed unaware of the PV diving on it. Then, the sub began a crash dive, but not in time. The Harpoon roared over it while the decks were still awash, and three MK47 depth bombs straddled the diving sub. LT Janes pulled around in a tight turn and watched the bombs explode, then saw an oil slick begin to form as debris and several cylindrical objects came to the surface. The cylinders were assessed as possible midget subs.

The routine of sector patrols was broken by a daily reconnaissance of Truk. This one-time "bastion of the Pacific" was now a wrecked and bypassed stronghold. Everything appeared to be in ruins...but A/A guns continued to give the PVs a hot time and many crews received their baptism by fire from the concealed emplacements along the bluffs and airstrips.

During the last four months of World War II several other squadrons flying PV-2s were deployed to the Pacific. VPB-144 had returned for a second tour, flying their Harpoons from Eniwetok on routine searches and patrols, as well as coordinating photo-recon missions against Ponape with Corsairs from MAG-13 and carrying out bombings of Wake. VPB-148 was back for a second tour operating from Midway and Johnston Islands, flying patrols in their Harpoons. VPB-153 based at Agana, Guam flew sector searches and ASW patrols in Harpoons before the war's end.

By the end of WWII, nearly half of the Navy's patrol squadrons were flying the Harpoon. The end of hostilities brought an accompanying reduction in the number of squadrons in the peacetime Navy and as a result, the PV-2 became the Navy's standard fleet patrol aircraft for the first three post-war years. As soon as design work had been completed on the PV-2, however, Wassal and his team set to work on the Harpoon's successor, the P2V Neptune. By 1947, PV-2 Harpoons were being replaced by the first Neptunes. The last Harpoons in the fleet were with VP-ML-3 until August 1948, but the aircraft continued to serve a long and useful life in eleven Navy Reserve Wings through the 1950s.

The only direct sales of Harpoons to a foreign government was to the Brazilian Navy in 1945. However, through the Foreign Military Assistance Programs (MAP) of the late 1940s, Portugal, Italy, France, Holland, South Africa and even Japan received surplus Harpoons and some of these are still flying today.

PV-2 Harpoon

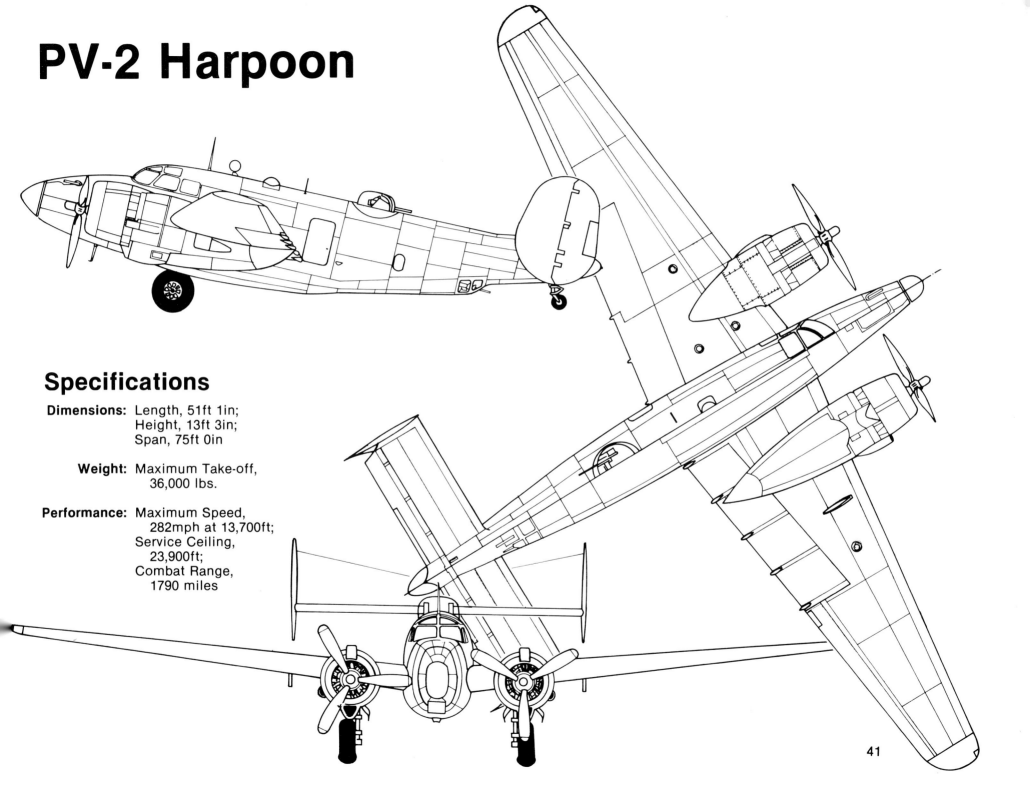

Specifications

Dimensions: Length, 51ft 1in; Height, 13ft 3in; Span, 75ft 0in

Weight: Maximum Take-off, 36,000 lbs.

Performance: Maximum Speed, 282mph at 13,700ft; Service Ceiling, 23,900ft; Combat Range, 1790 miles

The first PV-2 Harpoon in flight during early 1944. The straight wing and deepened oil cooler intake are visible. (Lockheed)

An early Harpoon over some rather spectacular scenery in California.

Four VPB-139 Harpoons return to Attu after a strike to the Kuriles in 1945. The P-38 coming up behind the Harpoons did not accompany them to the Kuriles, but served only as a 'Welcome Home' committee. (National Archives)

A rocket run by two of VPB-139's Harpoons is seen from the cockpit of a third PV-2. The target was a group of vessels in the cove. The Harpoons are pulling up after completing their runs. Rockets were usually fired in pairs or in a salvo (all eight at the same time) at a range of 1000 yards or less. (National Archives)

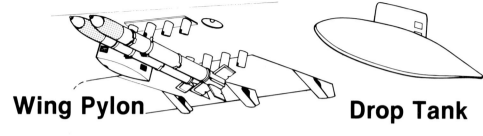

5" HVAR Rocket

Wing Pylon **Drop Tank**

(Left) A Hedron FAW-4 ordnance crew loads rockets on a VPB-139 harpoon at Attu. (National Archives)

(Bottom) One of the VPB-139's PV-2s is being refueled at Attu after a strike mission against the Kuriles during April 1945. (FAW-4 via L.A. Patteson)

A VPB-142 PV-2 in flight off Tinian. Note the navigator's head in the astro hatch (bubble) and since the turret was always manned on patrols a head is visible there also. (USN via Warren B. Herrick)

A late Harpoon in the immediate post war period at NAS Alameda, CA. The markings are in Yellow. The reason for the mirror-reverse of the underwing F62 isn't isn't known.

A Harpoon in overall Sea Blue is a rare but beautiful sight.

A rare bird - only 35 built - a PV-2D 'strafer'. Nose was upturned, pitot tubes moved to the top and eight .50 cal guns were mounted in a semi-circle under the nose. This BuNo 37543, is in immediate post-war Reserve colors, overall Sea Blue with Orange markings. (Bowers)

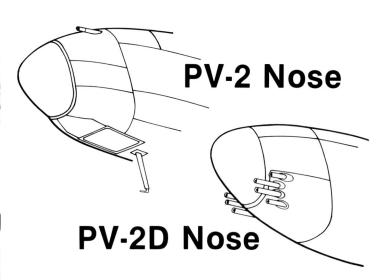

Several of the few PV-2Ds built were transferred to the Japanese, against whom they had been designed to fight. (National Archives)

Preliminary high altitude firing of a Tiny Tim rocket from PV-2, BuNo 37457, at the Naval Ordnance Test Station, Inyokern, CA. The white rectangles on the Harpoon's fuselage outline reference marks used in assessing the attitude of the rocket at ignition. Attitude was critical, as too much nose-down on the rocket would expose the firing aircraft tail structure to excessive rocket engine blast during ignition.

(Below Right) PV-2 bomb bay modifications permitted, among other things, two Tiny Tim rockets to be carried completely enclosed. The strut seen at the rear, between the rocket tail fins, assured a smooth drop of the first weapon. Rockets could only be fired singly, due to possible interference during initial free fall and powered flight.

P&W R-2800 engine

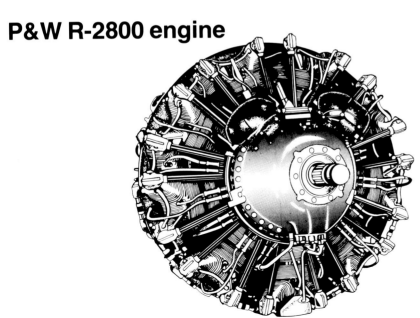

Harpoons served in Reserve squadrons through the 1950s. Here, a St. Louis-based PV-2 runs-up with the McDonnell plant in the backgound in 1955. Markings include White numbers, letters and nose tip. (Bowers)

A Harpoon flying the colors of the Brazilian Navy runs-up at Burbank in 1945. Lockheed-Vega manufactured B-17s, as in the background, under license from Boeing, as well as all Venturas and Harpoons. (Lockheed)

The experience gained from the Ventura/Harpoon enabled Lockheed to produce the next US Navy land-based maritime patrol bomber - the P2V Neptune. On 13 March 1947, this P2V-1 was delivered to VPML-2. (USN via A.A. Hoffman)

Naval Aviation
From
squadron/signal publications